Algebraic Structures and Operator Calculus

Mathematics and Its Applications

Managing Editor:

M. HAZEWINKEL

Centre for Mathematics and Computer Science, Amsterdam, The Netherlands

Also of interest:
Algebraic Structures and Operator Calculus. Volume I: Representations and Probability Theory, by P. Feinsilver and R. Schott, 1993, x + 224 pp., ISBN 0-7923-2116-2, MAIA 241.
Algebraic Structures and Operator Calculus. Volume II: Special Functions and Computer Science, by P. Feinsilver and R. Schott, 1994, x + 148 pp., ISBN 0-7923-2921-X, MAIA 292.

Volume 347

Algebraic Structures and Operator Calculus

Volume III:
Representations of Lie Groups

by

Philip Feinsilver

Department of Mathematics,
Southern Illinois University,
Carbondale, Illinois, U.S.A.

and

René Schott

CRIN,
Université de Nancy 1,
Vandoeuvre-les-Nancy, France

The MAPLE programming code pertaining to this book is available by anonymous ftp from: ftp.wkap.nl or by using your Web browser from URL: ftp://ftp.wkap.nl/software/algebraic_structures/lie-mapl.zip.

KLUWER ACADEMIC PUBLISHERS
DORDRECHT / BOSTON / LONDON

A C.I.P. Catalogue record for this book is available from the Library of Congress.

ISBN-13: 978-94-010-6557-3   e-ISBN-13: 978-94-009-0157-5
DOI: 10.1007/978-94-009-0157-5

Published by Kluwer Academic Publishers,
P.O. Box 17, 3300 AA Dordrecht, The Netherlands.

Kluwer Academic Publishers incorporates
the publishing programmes of
D. Reidel, Martinus Nijhoff, Dr W. Junk and MTP Press.

Sold and distributed in the U.S.A. and Canada
by Kluwer Academic Publishers,
101 Philip Drive, Norwell, MA 02061, U.S.A.

In all other countries, sold and distributed
by Kluwer Academic Publishers Group,
P.O. Box 322, 3300 AH Dordrecht, The Netherlands.

Printed on acid-free paper

To our families and friends

Table of Contents

Preface

The discovery of quantum theory showed that non-commutative structures are a fundamental feature of the physical world. One of the points we would like to emphasize is that the same is true in mathematics: non-commutative structures are natural and essential in mathematics. In fact, they explain many properties of mathematical objects — at a profound level.

In this work we answer some basic questions about Lie algebras and Lie groups.

1) Given a Lie algebra, for example, in matrix terms, or by prescribed commutation relations, how can one get a realization that gives some idea of what it 'looks like.' We present a concrete theory of representations with an emphasis on techniques suitable for (efficient) symbolic computing.

2) How do classical mathematical constructs interact with Lie structures? We take stochastic processes as an example. One can think of a Lie group as a non-commutative 'black box' and map functions, such as processes or trajectories of a dynamical system via the Lie algebra, and see what comes out.

Although some familiarity with groups is prerequisite, the basic methods used to find representations do not involve advanced knowledge. In fact, this is one of our main points. This book provides the reader techniques with which doing calculus on non-commutative structures can become as basic a tool as calculus on Euclidean space is presently.

The first author would like to express appreciation to Bruno Gruber for encouragement and support in many ways. He acknowledges the participants of the Carbondale Seminar over the past few years, with special thanks to J. Kocik and M. Giering. There is much appreciation to the Université de Nancy I: especially, to the Mathematics Seminar and to the computer science department, INRIA/Lorraine, for their support of this project over several years. Finally, the first author would like to acknowledge the hospitality and fruitful discussions during several visits in Toulouse, especially G. Letac and M. Casalis.

The second author thanks M. C. Haton for accepting to chair the department during the writing of this series. He is grateful to the Mathematics Department at SIU-C for kind hospitality during several visits.

We thank Randy Hughes for his TeX expertise that was so helpful for producing this series (including the Young tableaux for volume 2). Finally, we gratefully acknowledge the support of NATO, who provided travel and per diem support for several years.

Introduction

I. General remarks

Lie algebras and Lie groups play an increasingly prominent rôle in many applications of mathematics, notably in areas such as computer science, and control theory. They are essential in physics since the developments of relativity theory and quantum theory. Applications in computer science range from theoretical questions in computing and algorithm analysis (volume 2 of this series) and to practical situations such as robotic manipulation. Connections with probability theory are given in volume 1 of the series.

In this work we emphasize the original ideas of Lie algebras and Lie groups by taking a formal approach that works for Lie algebras in general. Although the geometric approach and the classification theory appear only implicitly, the details of our approach owe much to the great work of Cartan, Weyl, and Harish-Chandra.

First we present the models for Lie algebras that we will be using. It is important to see that Lie algebras appear quite naturally and are fundamental objects in mathematics. What we are doing here in this volume is providing the basic techniques of **calculus on Lie groups** that is needed as soon as one is encountering non-commutative structures.

1.1 SOME MODELS FOR LIE ALGEBRAS

The main models we will be using for Lie algebras are three:

1. Matrices
2. Vector fields
3. Shift operators on abstract states

1.1.1 Matrices

The fact that for matrices the product is noncommutative: $AB \neq BA$ in general, is a basic feature of matrix algebra. All of our Lie algebras will be finite-dimensional and have faithful realizations as matrices. This is the import of *Ado's Theorem*.

The basic things we will be considering may be illustrated by the following example. Consider matrices of the form

$$X = \begin{pmatrix} a_1 & a_2 \\ 0 & a_3 \end{pmatrix} = a_1 \xi_1 + a_2 \xi_2 + a_3 \xi_3$$

with $a_i \in \mathbf{C}$. The fact that the vector space $\mathcal{V}$, generated by $\{\xi_i\}$ forms a Lie algebra is just the fact that the commutators $[\xi_i, \xi_j] = \xi_i \xi_j - \xi_j \xi_i$ belong to $\mathcal{V}$, for $1 \leq i, j \leq 3$. Here we have the commutation relations

$$[\xi_1, \xi_2] = \xi_2, \quad [\xi_2, \xi_3] = \xi_2, \quad [\xi_1, \xi_3] = 0$$

The important observation of Lie was that if the matrices form a Lie algebra, then the exponentials e^X form a group, the inverse to e^X given by e^{-X}. A basic question is how to find *representations* of the Lie algebra and corresponding Lie group as matrices, or, more generally, as linear operators on some vector space.

1.1.2 Vector fields

By vector fields, we mean first-order partial differential operators acting on C^∞ functions on some domain in Euclidean space. The composition of these operators forms second-order operators (in general). However, the commutator remains a first-order operator. (See §3.5 below for more discussion.)

We will show in Chapter 2 how to realize a Lie algebra as vector fields. They arise naturally as the derivatives of the action of Lie groups on spaces of functions. For example, consider one of the simplest non-trivial Lie structures: the three-dimensional Lie algebra with basis ξ_1, ξ_2, ξ_3 satisfying the commutation relations $[\xi_3, \xi_1] = \xi_2$, with ξ_2 commuting with the other two basis elements. This is the *Heisenberg algebra* that provides a basic example for illustrating many features of non-commutative structures. One realization as vector fields is given by setting $\xi_1 = x\partial/\partial y$, $\xi_2 = \partial/\partial y$, $\xi_3 = \partial/\partial x$. In many cases, we will find representations on the space of polynomials. Notice that this particular realization illustrates this feature.

1.1.3 Shift operators

One can consider abstract vectors or states of some system, such as the files in a computer or quantum states describing a physical system. Denoting these by $[\![n]\!]$, we have shift operators acting on them. For integer $n \geq 0$, we have $\mathcal{R}$, the *raising operator* which acts according to $\mathcal{R}[\![n]\!] = [\![n+1]\!]$. The operator $\mathcal{V}$ defined by the action $\mathcal{V}[\![n]\!] = n[\![n-1]\!]$ is a formal differentiation operator and satisfies the Lie bracket (commutator) relation $[\mathcal{V}, \mathcal{R}] = I$, the identity operator. Again, this is a realization of the Heisenberg algebra. An important feature of this example is that the basis for the vector space is generated by the action of the algebra, in fact of $\mathcal{R}$, on the single vector $[\![0]\!]$: $[\![n]\!] = \mathcal{R}^n[\![0]\!]$. Thus $[\![0]\!]$ is called a *cyclic vector* or, since it is the lowest or zero state, and is annihilated by $\mathcal{V}$, the *vacuum state*.

1.2 OVERVIEW OF THIS VOLUME

After this introductory chapter with notations and basic definitions concerning Lie algebras that we will use, the next four chapters present our approach to representations of Lie algebras and Lie groups. This involves first discussing the *boson calculus* which is a general version of the $\mathcal{R}$ and $\mathcal{V}$ calculus alluded to in the previous paragraph. Next we show how to realize a given Lie algebra as vector fields and in terms of generalized differential operators. In Chapters 3 and 4 we present our theory of "polynomial representations" of Lie groups. Chapters 5 through 7 discuss specific examples or classes of Lie algebras and corresponding groups studied using the theory developed in Chapters 1–4. The work continues with Chapter 8 concerning properties of the functions that arise as *special functions* – i.e., functions having special properties, here because of the connection with representations of Lie algebras. And we conclude with some computational aspects that have been recently developed thanks to the advent of software packages such as MAPLE. In fact, in parallel with examples throughout the work, we have MAPLE output of the routines we have developed collected in a special section at the end of the book. The important feature is that expressing things in terms of matrices makes symbolic computations feasible.

Now we present some details of our "polynomial representations." We use the quotes since in general, the basis functions will not be polynomial in all variables. The guiding idea is solving evolution equations with polynomial initial conditions:

$$\frac{\partial u}{\partial t} = X u, \qquad u(0) = f$$

where f is a polynomial in underlying variables. A natural technique to use is that of generating functions, namely, to solve

$$\frac{\partial u}{\partial t} = X u, \qquad u(0) = g(A)$$

where A indicates some parameter variables and $g(A)$ is an exponential in the original variables. Then expanding in A gives the solutions for polynomials. The idea in this work is to use not only the usual generating functions, but *non-commutative* generating functions, i.e., generating functions in non-commuting variables. In such a case, X is an element of a Lie algebra, realized either as a matrix or as a vector field, i.e., a first-order differential operator, and $g(A)$ is a group element near the identity. The basic principle of Lie is very useful here, namely, that systems having the same Lie structure, commutation relations, have isomorphic flows.

First, consider $D = d/dx$, and

$$\frac{\partial u}{\partial t} = D u, \qquad u(0) = e^{Ax}$$

with solution $u = e^{tD} e^{Ax} = e^{Ax+At} = \sum (A^n/n!)(x+t)^n$. And the corresponding solutions to $u_t = Du$, $u(0) = x^n$ are $(x+t)^n$. The next main step is to go to more general operators. So

$$\frac{\partial u}{\partial t} = H(D)u, \qquad u(0) = e^{Ax} \tag{1.2.1}$$

Now we have
$$u = e^{tH(D)} e^{Ax} = e^{Ax+tH(A)}$$

where we assume, e.g., that H is an analytic function in a neighborhood of $0 \in \mathbf{C}$. Now, if we can write
$$e^{tH(A)} = \int_{-\infty}^{\infty} e^{Ay} p_t(dy) .$$

for some measures p_t, which we typically take to be probability measures, we can use the solution to $u_t = Du$ to solve the more general case, thus:

$$e^{tH(D)} e^{Ax} = \int_{-\infty}^{\infty} e^{yD} p_t(dy) e^{Ax} = \int_{-\infty}^{\infty} e^{Ax+Ay} p_t(dy)$$

$$= \sum_{n \geq 0} \frac{A^n}{n!} \int_{-\infty}^{\infty} (x+y)^n p_t(dy)$$

Thus, the *moment polynomials:*

$$h_n(x,t) = \int_{-\infty}^{\infty} (x+y)^n p_t(dy)$$

are solutions to (1.2.1) with polynomial initial conditions and yield polynomial solutions. The main feature here is the product of the exponentials: $e^{yD} e^{Ax}$, which is the product of group elements and is thus itself an element of the Heisenberg group. This is where the Lie structure comes in. In general, the Lie algebra with basis $\{1, X_i, D_j\}$, where the variables X_i act as multiplication by x_i and the D_j act as partial derivatives $\partial/\partial x_j$, we call the standard Heisenberg algebra and the associative algebra generated (say as operators on polynomials or smooth functions) is the *Heisenberg-Weyl algebra.* The general properties of such systems, i.e., isomorphic as Lie structures, is the *boson operator calculus.* The moment systems indicated above generalize directly to several variables using the boson operator calculus. We will call the corresponding moment systems and associated calculus *canonical Appell systems.* In general, we have these principal aspects:

1. An underlying Lie structure is given at the beginning.

2. Everything is expressed in terms of the boson operator calculus.

One extension of the above constructions is as follows. Given a Lie algebra, we want to solve
$$\frac{\partial u}{\partial t} = H(\xi)u , \qquad u(0) = g(A)$$

where ξ is a basis of the given algebra and g denotes a typical element of the group. Then, in analogy to the above calculation, with the exponential map denoted by the usual exponential:

$$u = e^{tH(\xi)} g(A) = \int e^{Y\xi} p_t(dY) g(A)$$

$$= \int g(Y \odot A) p_t(dY)$$

where $\odot$ denotes the composition law of the associated Lie group. This is a convolution integral, and we can interpret the associated family of measures p_t as corresponding to a process with independent increments on the Lie group. Expanding in terms of the non-commutative variables ξ, we find a general moment system. In this general situation, we call these *Appell systems on Lie groups*. Operator calculus on the Lie group is used to study convolution structures and stochastic processes on the group. The form of the stochastic processes on the group shows clearly the interaction of the process with the non-commutative nature of the Lie system.

Now we can explain the meaning of the title of this series. Operator calculus is calculus for the associative algebras and Lie groups generated by Lie algebras, using the tool of boson operator calculus and studying connections with convolution structures and stochastic processes.

II. Notations

For indices, m, n will be used for multi-indices, i, j, k, l always single. For $n = (n_1, \ldots, n_N)$, $n! = n_1! \cdots n_N!$, $|n| = \sum n_i$, $n \geq 0$ means $n_i \geq 0$, $\forall i$. Multi-index binomial coefficients are defined as

$$\binom{n}{m} = \prod_i \binom{n_i}{m_i}$$

The standard basis vectors are e_i, with $e_1 = [1, 0, 0, \ldots]$, $e_2 = [0, 1, 0, \ldots]$, etc. Thus, $n = \sum_i n_i e_i$. For N-dimensional space, the variables are denoted $x = (x_1, \ldots, x_N)$. The partial derivative with respect to x_i will be denoted by D_i, with $D = (D_1, \ldots, D_N)$. For the group coordinates, we have A_i, $\partial_j = \partial/\partial A_j$.

In general, any array will be indicated by the corresponding single letter, e.g., $A = (A_1, \ldots, A_d)$, $f(A)$ denotes a function of $A_1, \ldots, A_d$. If A and B are two arrays, then AB will denote $\sum A_i B_i$.

Repeated Greek indices are always summed, regardless of position. E.g., $a_\mu b_\mu = \sum_i a_i b_i$. The variables s, t will always denote single variables, i.e., single parameters. Differentiation with respect to a single parameter is denoted by a dot, e.g. $\dot{A}$.

Abstract states will be denoted by $[\![n]\!]$, with n in general a multi-index. Matrix elements of operators acting on the abstract states will be denoted $\left\langle \begin{smallmatrix} m \\ n \end{smallmatrix} \right\rangle$, for example, for the transition from state $[\![n]\!]$ to $[\![m]\!]$.

Variables (other than indices, of course) are assumed to be complex-valued unless indicated otherwise. The main restriction is that all probability measures are assumed to have support in $\mathbf{R}$ or in some $\mathbf{R}^N$.

Expected value with respect to an implicit probability measure is denoted by angle brackets. E.g.,

$$\langle f(X) \rangle = \int_{-\infty}^{\infty} f(x)\, p(dx)$$

The Kronecker delta is denoted δ_{ij}: 1 if $i = j$, 0 otherwise.

The symbol θ_{ij} denotes the *ordering symbol* equal to 1 if $i < j$, zero otherwise.

2.1 BOSON CALCULUS

The standard example of boson variables are X_i, D_j, where X_i is the operator of multiplication by x_i.

2.1.1 Definition. Define operators $\mathcal{R}_j$, $\mathcal{V}_j$, on $[\![n]\!]$, thus:

$$\mathcal{R}_j[\![n]\!] = [\![n + e_j]\!], \qquad \mathcal{V}_j[\![n]\!] = n_j [\![n - e_j]\!]$$

They satisfy the Heisenberg algebra (boson algebra) commutation relations $[\mathcal{V}_k, \mathcal{R}_j] = \delta_{jk} I$. We also have

2.1.2 Definition. The boson operators $\bar{\mathcal{R}}_j$, $\bar{\mathcal{V}}_j$ are defined by

$$\bar{\mathcal{R}}_j[\![n]\!] = (n_j + 1)[\![n + e_j]\!], \qquad \bar{\mathcal{V}}_j[\![n]\!] = [\![n - e_j]\!]$$

2.2 APPELL SYSTEMS

For Appell systems:

1. H denotes the generator of the time-evolution, with corresponding Appell polynomials denoted h_n.

2. The canonical variables are denoted by y_i and V_i, with corresponding Appell polynomials η_n.

3. The raising operators are R_i. The lowering operators V_i are functions of z, corresponding to D via the exponential e^{zx}.

2.3 LIE ALGEBRAS

$\mathcal{G}$ denotes a generic Lie algebra of dimension d. The basis is denoted by $\{\xi_i\}$. A typical element of the algebra is $X = \alpha_\mu \xi_\mu$.

left dual	$\xi^\ddagger = \pi^\ddagger \partial$	double dual	$\hat{\xi} = \mathcal{R}\hat{\pi}(\mathcal{V})$
right dual	$\xi^* = \pi^* \partial$	double right dual	$\hat{\xi}^* = \mathcal{R}\hat{\pi}^*(\mathcal{V})$

For the enveloping algebra we have

basis $\quad [\![n]\!] = \xi^n = \xi_1^{n_1} \cdots \xi_d^{n_d} \quad$ coordinates $\quad c_n(A) = A^n/n! = A_1^{n_1} \cdots A_d^{n_d}/(n_1! \cdots n_d!)$

The discrete dual, $\bar{\xi}$, acts on the index m of the matrix element $\left\langle \begin{smallmatrix} m \\ n \end{smallmatrix} \right\rangle$ according to

$$\sum_m \bar{\xi} \left\langle \begin{matrix} m \\ n \end{matrix} \right\rangle [\![m]\!] = \sum_m \left\langle \begin{matrix} m \\ n \end{matrix} \right\rangle \xi[\![m]\!]$$

Denote the (matrix form of the) adjoint representation by $\check{\xi}$. The exponential of the adjoint representation: $\check{\pi} = \hat{\pi}^{-1}\hat{\pi}^*$. The *splitting lemma* reads $\dot{A} = \alpha\pi^*(A) = \alpha\pi^\ddagger(A)$.

Remark. Because of the various duals noted above, we will use † for the usual matrix transpose.

2.4 HOMOGENEOUS SPACES

The Cartan splitting is denoted $\mathcal{G} = \mathcal{P} \oplus \mathcal{K} \oplus \mathcal{L}$ at the algebra level, with corresponding factorization of the group $\mathbf{G} = \mathbf{PKL}$. (Note: the $\mathcal{P}$ and $\mathbf{P}$ denote capital Greek rho's.) The corresponding block matrix form is

$$X = \begin{pmatrix} \alpha & v \\ -\beta & \delta \end{pmatrix}$$

and the group factorization

$$g = \begin{pmatrix} I & V \\ 0 & I \end{pmatrix} \begin{pmatrix} E & 0 \\ 0 & D \end{pmatrix} \begin{pmatrix} I & 0 \\ -B & I \end{pmatrix} \tag{2.4.1}$$

The basis of the Lie algebra is given by R_i for $\mathcal{P}$, with coordinates v_i and V_i for the algebra and group respectively. For $\mathcal{L}$: basis L_i, coordinates β_i, B_i. For $\mathcal{K}$, we have the Cartan subalgebra with basis ρ_i, with coordinates for the Cartan elements h_i and H_i and the roots ρ_{ij}, with coordinates κ_{ij}, C_{ij}. For the special element ρ_0 we have coordinates h_0 and H_0. In the general discussion, coordinates for $\mathcal{K}$ will be denoted by A, corresponding to the block diagonal matrix, with blocks E, D, in equation (2.4.1). Note that this gives group elements

$$g = g(V, A, B) = g(V, H, C, B)$$

depending on the context.

Matrix elements for the action of the group on the basis for the induced representation $\psi_n = R^n \Omega$ are denoted $\left\langle\!\left\langle {m \atop n} \right\rangle\!\right\rangle$. The squared norms are $\langle \psi_n, \psi_n \rangle = \gamma_n$.

For coherent states: $\psi_V = e^{VR} \Omega$ with inner product $\langle \psi_B, \psi_V \rangle = \Upsilon_{BV}$.

2.5 TERMINOLOGY OF LIE ALGEBRAS

GL(n) denotes $n \times n$ matrices with non-zero determinant.

O(n) denotes the orthogonal group of matrices M satisfying $M^\dagger M = I$.

U(n) denotes the group of $n \times n$ complex unitary matrices (i.e., satisfying $\bar{M}^\dagger M = I$).

SO(n) denotes orthogonal matrices with determinant equal to one.

so(n) is the corresponding Lie algebra of skew-symmetric matrices, $M^\dagger = -M$.

Sp($2n$) is the symplectic group of dimension $2n$, preserving an alternating bilinear form, e.g., $M^\dagger J M = J$, with $J = \left({O \atop -I} {I \atop O} \right)$ in block form.

sp($2n$) is the corresponding Lie algebra.

SU(n) is the group of unitary matrices of determinant one.

su(n) is the Lie algebra of $n \times n$ skew-Hermitian matrices (entries in $\mathbf{C}$).

sl(n) is the Lie algebra of $n \times n$ matrices of trace zero corresponding to the Lie group SL(n) of matrices of determinant one.

Affine groups, $Ax + b$, with $A \in$ GL(d), $b \in \mathbf{R}^d$, acting on $x \in \mathbf{R}^d$, are denoted Aff$d$, with corresponding Lie algebra aff$d$, e.g., Aff2 and aff2.

Euclidean groups, $Ax + b$, with $A \in$ SO(d), $b \in \mathbf{R}^d$, acting on $x \in \mathbf{R}^d$, are denoted Ed, with corresponding Lie algebra ed, e.g., E5 and e5.

The group of $d \times d$ (strictly) upper-triangular matrices is denoted Nd with Lie algebra nd. E.g., N5 and n5.

III. Lie algebras: some basics

Here we review some basic facts and features about Lie algebras and associated Lie groups.

3.1 LIE ALGEBRAS

The abstract definition of a Lie algebra is as follows. A Lie algebra is a vector space with a non-associative multiplication — the Lie bracket, denoted $[X, Y]$ — satisfying:

$$[X, X] = 0$$
$$[X, [Y, Z]] + [Y, [Z, X]] + [Z, [X, Y]] = 0$$

The first condition implies antisymmetry $[X, Y] = -[Y, X]$ (as we are always working over subfields of $\mathbf{C}$) and the second condition is the *Jacobi identity*.

A subspace of an associative algebra is a Lie algebra if it is closed under the Lie bracket given by the commutator $[X, Y] = XY - YX$. In our work we will always use this form of the Lie bracket.

A *representation* of a Lie algebra is a realization of the Lie algebra as linear operators on a vector space with the Lie bracket given by the commutator. If the vector space is a Hilbert space and the group elements are unitary operators, then one has a *unitary representation*.

3.2 ADJOINT REPRESENTATION

Given a basis $\{\xi_1, \xi_2, \ldots, \xi_d\}$ for a Lie algebra $\mathcal{G}$ of dimension d, the Lie brackets

$$(\operatorname{ad}\xi_k)(\xi_j) = [\xi_k, \xi_j] = \sum_i c^i_{kj} \xi_i$$

define the *adjoint representation*, given by the matrices $\operatorname{ad}\xi_k$, which we denote by $\breve{\xi}_k$

$$(\breve{\xi}_k)_{ij} = c^i_{kj}$$

The fact that this is a representation follows from the Jacobi identity.

A subalgebra $\mathcal{A}$ is *abelian* if $[R, S] = 0$ for all $R, S \in \mathcal{A}$. An *ideal* $\mathcal{I}$ in $\mathcal{G}$ is a non-zero subalgebra such that $[X, Z] \in \mathcal{I}$ for all $X \in \mathcal{G}$, $Z \in \mathcal{I}$. That is, $\mathcal{G}$ acts on $\mathcal{I}$ by the adjoint representation. In general, if $\mathcal{A}$ and $\mathcal{B}$ are two subalgebras of $\mathcal{G}$ such that $A \in \mathcal{A}, B \in \mathcal{B}$ implies $[A, B] \in \mathcal{B}$, then we say that $\mathcal{A}$ *normalizes* $\mathcal{B}$. The *center* of $\mathcal{G}$ consisting of all elements Z such that $[X, Z] = 0$ for all $X \in \mathcal{G}$, is an abelian ideal if it is non-trivial.

A Lie algebra is *semisimple* if it has no abelian ideals.

3.2.1 Derived and central series

The linear span of commutators $[X, Y]$, $X, Y \in \mathcal{G}$, is a Lie subalgebra of $\mathcal{G}$, denoted $\mathcal{G}'$, the *derived algebra*. The *derived series* is given by the sequence of ideals

$$\mathcal{G} = \mathcal{G}^{(0)} \supset \mathcal{G}' \supset \cdots \supset \mathcal{G}^{(j)} \supset \cdots$$

where $\mathcal{G}^{(j+1)} = (\mathcal{G}^{(j)})'$. $\mathcal{G}$ is *solvable* if for some r, $\mathcal{G}^{(r)} = 0$.

The *central series* is given by $\mathcal{G}_{(0)} = \mathcal{G}$, $\mathcal{G}_{(j)} \supset \mathcal{G}_{(j+1)} = [\mathcal{G}, \mathcal{G}_{(j)}]$, with the bracket of subspaces denoting the subspace generated by brackets of their elements. If for some r, $\mathcal{G}_{(r)} = 0$, then $\mathcal{G}$ is *nilpotent* and the least index r having this property is called the *step* of the algebra. Notice that if $\mathcal{G}$ is a step-r nilpotent Lie algebra, then $\mathcal{G}_{(r-1)}$ is an abelian ideal, in fact it is in the center of $\mathcal{G}$.

3.2.2 Cartan's criteria

The *Killing form* is a bilinear form on $\mathcal{G} \times \mathcal{G}$ defined by $\kappa(X, Y) = \operatorname{tr}((\operatorname{ad} X)(\operatorname{ad} Y))$, i.e., in the notation above, $\operatorname{tr}(\check{X}\check{Y})$. Then Cartan's criteria are:

1. $\mathcal{G}$ is nilpotent if $\kappa(X, Y) = 0$ for all $X, Y \in \mathcal{G}$.

2. $\mathcal{G}$ is solvable if $\kappa(X, Y) = 0$ for all $X \in \mathcal{G}$, $Y \in \mathcal{G}'$.

3. $\mathcal{G}$ is semisimple if $\kappa(X, Y)$ is a non-degenerate form on $\mathcal{G} \times \mathcal{G}$.

Note that $\mathcal{G}$ is solvable if and only if $\mathcal{G}'$ is nilpotent.

The *Levi-Mal'cev Theorem* says that any Lie algebra is of the form $\mathcal{K} \oplus \mathcal{B}$ where $\mathcal{K}$ is semisimple and $\mathcal{B}$ is a maximal solvable ideal. We will say that a Lie algebra is of *affine type* if it has an abelian ideal. In that case, we let the ideal $\mathcal{P}$ be maximal abelian and write $\mathcal{G} = \mathcal{P} \oplus \mathcal{K}$. Note that $\mathcal{K}$ normalizes $\mathcal{P}$, i.e., the adjoint action of $\mathcal{K}$ on $\mathcal{P}$ gives a representation of $\mathcal{K}$.

3.3 UNIVERSAL ENVELOPING ALGEBRA

Given a representation of a Lie algebra $\mathcal{G}$ in terms of matrices (more generally, as operators on some linear space), the *enveloping algebra* is the associative algebra generated by these operators within the algebra of linear operators acting on the given representation space. The universal enveloping algebra $\mathcal{U}(\mathcal{G})$ is an abstract construction of an associative algebra generated by the Lie algebra such that the Lie bracket becomes the commutator. With $\{\xi_1, \ldots, \xi_d\}$ a basis for $\mathcal{G}$, ξ denoting the d-tuple $(\xi_1, \ldots, \xi_d)$, n the multi-index $(n_1, \ldots, n_d)$, define the monomials

$$[\![n]\!] = \xi^n = \xi_1^{n_1} \cdots \xi_d^{n_d}$$

Note that the products involving ξ_j are ordered. The Poincaré-Birkhoff-Witt theorem is to the effect that the $[\![n]\!]$ form a basis for the universal enveloping algebra.

The basic idea behind our approach is to find the left and right regular representations for the basis elements ξ_j acting on the universal enveloping algebra. On any associative algebra, $\mathcal{M}$, say, we can define mappings from $\mathcal{M}$ into $\mathrm{Lin}(\mathcal{M})$, linear mappings on $\mathcal{M}$, by

$$r_A X = XA, \quad l_A X = AX$$

Notice that l gives a representation, i.e., it is a homomorphism, of $\mathcal{M}$: $l_{AB} = l_A l_B$, the order of multiplication is preserved; while r is an anti-homomorphism: $r_{AB} = r_B r_A$, the order of multiplication is reversed. The representation l is called the *left regular representation* of $\mathcal{M}$.

3.3.1 Proposition.   *For all $A, B \in \mathcal{M}$, l_A and r_B commute. I.e., $[l_A, r_B] = 0$.*

Notice that this is exactly the associativity of multiplication: $(AX)B = A(XB)$. This has the following consequence

3.3.2 Proposition.   *For given $A, B \in \mathcal{M}$, define the operator Θ on $\mathcal{M}$ by $\Theta(X) = AX + XB$. Then*

$$\Theta^n(X) = \sum_{k=0}^{n} \binom{n}{k} (A^{n-k} X B^k)$$

   Proof: This follows by induction as in the proof of the binomial theorem, using Proposition 3.3.1.   ∎

3.4 KIRILLOV FORM

The Kirillov form, K, is an antisymmetric form, linear in variables $\{x_i\}$, given in terms of the adjoint representation by

$$K(x)_{ij} = x_\mu c_{ij}^\mu$$

We can write the commutation relations in terms of K as

$$[\xi_i, \xi_j] = K(\xi)_{ij}$$

Note that the ith row of $K(x)$ is given by $x\check{\xi}_i$, where x is the vector $(x_1, \ldots, x_d)$.

The Lie bracket of two elements $Y = \alpha_\mu \xi_\mu$, $X = x_\mu \xi_\mu$ has the form

$$[Y, X] = \langle \alpha, K(\xi)x \rangle$$

where the brackets denote the usual dot product of vectors (here x is treated as a column vector). Consequently, the center of $\mathcal{G}$ is determined by the null-space of $K(x)$ that is independent of the values of the x_i. I.e., $\alpha K(x) = 0$, for all values x_i determines $\alpha_\mu \xi_\mu$ in the center.

3.4.1 Change-of-basis

Consider the change-of-basis $\eta_i = \xi_\mu T_{\mu i}$, with

$$[\eta_i, \eta_j] = \bar{c}^\mu_{ij}\eta_\mu = \bar{K}(\eta)_{ij}$$

Or,

$$[\xi_\lambda T_{\lambda i}, \xi_\mu T_{\mu j}] = \bar{K}(\eta)_{ij}$$

I.e.,

$$T^\dagger K(\xi) T = \bar{K}(\eta)$$

showing that K transforms as a bilinear form, as expected.

3.5 VECTOR FIELDS AND FLOWS

Vector fields are first-order partial differential operators:

$$X = a_\mu D_\mu$$

with a_μ functions of underlying variables x_i, $D_i = \partial/\partial x_i$. They form a Lie algebra with the product of $X = a_\mu D_\mu$, $Y = b_\mu D_\mu$ given by

$$[X, Y] = (Xb_\mu - Ya_\mu)D_\mu = \big(a_\mu(D_\mu b_\nu) - b_\mu(D_\mu a_\nu)\big)D_\nu$$

The flow corresponding to $X = a_\mu D_\mu$ is given by solutions to the differential equation

$$\frac{\partial u}{\partial t} = a_\mu(x)D_\mu u \tag{3.5.1}$$

Any function u of the form $u(x,t) = f(x(t))$ where $x(t)$ satisfies the ordinary differential system

$$\dot{x}_i(t) = a_i(x(t))$$

with $x_i(0) = x_i$ is a solution of (3.5.1). This is the *method of characteristics*. The point is that you only have to solve for the identity function as initial condition, i.e., $x(t)$ is the solution to (3.5.1) with initial condition $u(x,0) = x$, and the solution to (3.5.1) with initial condition $u(x,0) = f(x)$ is $f(x(t))$.

3.6 JORDAN MAP

A Lie algebra of matrices can be written as vector fields with linear coefficients by the *Jordan map*:

$$X = (X_{ij}) \leftrightarrow x_\lambda X_{\lambda\mu} D_\mu$$

The correspondence of Lie brackets is the relation

$$x_\lambda [X, Y]_{\lambda\mu} D_\mu = [x_\lambda X_{\lambda\mu} D_\mu, x_\alpha Y_{\alpha\beta} D_\beta]$$

where the bracket on the left is the commutator of matrices and on the right the Lie bracket of vector fields.

Remark. Generally, we will write the Jordan map using the abstract boson operators $\mathcal{R}, \mathcal{V}$, corresponding respectively: $x_i \leftrightarrow \mathcal{R}_i$, $D_i \leftrightarrow \mathcal{V}_i$. Thus, $\mathcal{R}_i \mathcal{V}_j$ corresponds to the elementary matrix with ij entry equal to 1, all other entries zero. So we have

$$X = (X_{ij}) \leftrightarrow \mathcal{R}_\lambda X_{\lambda\mu} \mathcal{V}_\mu$$

for the Jordan map in boson variables.

3.7 LIE GROUPS

Corresponding to elements of a Lie algebra are group elements via exponentiation. Given A, a matrix, the series

$$e^{tA} = \sum_{n \geq 0} \frac{t^n A^n}{n!}$$

is well defined. For $-\infty < t < \infty$, $U(t) = e^{tA}$, comprises a group — the *one-parameter subgroup* generated by A. They satisfy

$$U(t + s) = U(t)U(s), \qquad U(0) = I \quad \text{identity operator}$$

$$\frac{dU}{dt} = AU = UA$$

Note that the inverse of $U(t)$ is $U(-t)$. By using an operator norm, e.g. for matrices or bounded operators, say on a Banach space, $\|A\| = \sup \|Av\|/\|v\|$, $v \neq 0$, these equations hold in the algebra of bounded operators. There are two further natural interpretations to see:

1) In terms of a representation, i.e., acting on a fixed vector v, we have the vector-valued function:

$$u(t) = e^{tA} v, \qquad u(0) = v$$

And $du/dt = Au$. Conversely, for finite-dimensional spaces one can compute explicitly the solution to $du/dt = Au$, $u(0) = v$ as $\exp(tA)\,v$.

2) On an inner product space, (pre-)Hilbert space, one considers the matrix elements

$$u(t) = \langle v, e^{tA}\,w \rangle$$

for fixed vectors v, w. Then $u(t)$ is a scalar function satisfying

$$\frac{du}{dt} = \langle v, Ae^{tA}w \rangle, \qquad u(0) = \langle v, w \rangle$$

As for case 1), on any vector space, not necessarily finite-dimensional, the operator A acts nilpotently if, for any given vector v, for all n sufficiently large, $A^n v = 0$. In such a case, both 1) and 2) make sense since the exponentials can be defined by finite series.

One can exponentiate Prop. 3.3.2, multiplying both sides by $t^n/n!$ and summing:

$$e^{t\Theta}(X) = e^{tA}\,X e^{tB}$$

For the case when $B = -A$ we have:

3.7.1 Proposition.

$$e^{tA}\,X e^{-tA} = e^{t\,\mathrm{ad}\,A}\,X$$

3.8 ADJOINT GROUP AND COADJOINT ORBITS

Let Y be a vector field $a_\mu D_\mu$. Denote by Ω the constant function equal to 1. Using the fact that $Y\Omega = 0$, we can write the solution to (3.5.1) in operator form as follows:

$$u(x,t) = e^{tY}\,f(x) = e^{tY}\,f(\bar{X})e^{-tY}\,\Omega$$

where $\bar{X}$ denotes the operator of multiplication by x. The idea is that for polynomial f, say, we have the operator relations

$$U(t,\bar{X}) = e^{tY}\,f(\bar{X})e^{-tY} = f(e^{tY}\,\bar{X}e^{-tY})$$

That is, with $\bar{X}(t) = e^{tY}\,\bar{X}e^{-tY}$, we have $u(x,t) = f(\bar{X}(t))1$, an operator version of the method of characteristics. The operator $U(\bar{X},t)$ satisfies the Heisenberg equations of motion

$$\dot{U} = [Y,U]$$

equivalently, $\dot{U} = (\mathrm{ad}\,Y)U$.

Given the Lie algebra $\mathcal{G}$, the group generated by the matrices $\check{\xi}$ of the adjoint representation is the *adjoint group*.   From Proposition 3.7.1 we have

$$e^{\check{Y}}[X]_\xi = [e^Y X e^{-Y}]_\xi$$

i.e., the exponential of the adjoint representation on the Lie algebra is the action of the group by conjugation on the algebra. In this equation, the left-hand side is calculated in terms of the matrices of the adjoint representation acting on a vector of coordinates, $[X]_\xi$. On the right-hand side, say with $\mathcal{G}$ given as matrices, the calculation is done directly with the realization of $\mathcal{G}$ as given, then written as a vector of coordinates with respect to the basis ξ. For example, if $\mathcal{G}$ is so(n), of dimension $d = \binom{n}{2}$, on the left side the matrices are $d \times d$, while on the right, they are $n \times n$.

Now consider the flow on $\mathcal{G}$ generated by an element $Y = \alpha_\mu \xi_\mu$ corresponding to the adjoint action

$$\dot{X} = [Y, X] \tag{3.8.1}$$

so that $X(t) = e^{tY} X(0) e^{-tY}$. With $X(0) = x_\mu \xi_\mu$, we have $X(t) = x(t)_\mu \xi_\mu$. This can be written as the system

$$\dot{x}(t)_i = \alpha_\lambda c^i_{\lambda\mu} x_\mu(t) \tag{3.8.2}$$

or, more succinctly, in terms of the Kirillov form

$$\dot{X} = \langle \alpha, K(\xi)x \rangle$$

The solution to eq. (3.8.2) is given in terms of the matrices of the adjoint representation as

$$x(t) = e^{t\check{Y}} x$$

where we treat the coordinates as a column vector.

Note that the vector field generating the flow in (3.8.2) is

$$\alpha_\lambda c^\nu_{\lambda\mu} x_\mu D_\nu = \alpha_\lambda (\check{\xi}^\dagger_\lambda)_{\mu\nu} x_\mu D_\nu$$

($\dagger$ denoting transpose) i.e., the vector fields corresponding to the Lie algebra are given by the Jordan map of the transposed adjoint representation. At $t = 1$, the coordinates α parametrize the manifold of solutions, the *adjoint orbits*.

Going back to eq. (3.8.1), look at the flow directly on $\mathcal{G}$. For each basis element ξ_i, we have $\dot{\xi}_i = [Y, \xi_i]$. We treat the array $\xi = (\xi_1, \ldots, \xi_d)$ as a row vector and write

$$\dot{\xi}_i = \alpha_\lambda c^\mu_{\lambda i} \xi_\mu = \alpha_\lambda K_{\lambda i}(\xi)$$

in the form $\dot{\xi} = \xi \check{Y}$. Thus,

$$\xi(t) = \xi e^{t\check{Y}}$$

Evaluating at $t = 1$, we have $\xi(1)$ as a function of the group parameters α. These are the *coadjoint orbits*.

Given an initial $X(0)$, evaluating at $t = 1$, the manifold of solutions $X(1)$ corresponding to all Y gives the homogeneous space G/H where G denotes the group generated by the adjoint representation and H, the isotropy subgroup fixing $X(0)$, is generated by Y such that $[Y, X(0)] = 0$.

References

General mathematical references for Lie theory are Bourbaki [6] and Chevalley [8]. Biedenharn & Louck [3], Klimyk & Vilenkin [40] present Lie theory from the physicists' point-of-view. References for geometric quantization and connections with symmetric spaces are Woodhouse [56] and Hurt [36].

Here are some examples where Lie theory is involved:

- Control theory: Cheng, Dayawansa & Martin [7], Crouch & Irving [9], Hermes [33], Jakubczyk & Sontag [38], Sussmann [49].

- Robotics: Murray, Li, & Sastry [46].

- Computer science: Duchamp & Krob [11], [12], and volume 2 [18].

- Combinatorics perspective: Viennot [51].

For computations with MAPLE, see, e.g., Koornwinder [42].

Chapter 1   Operator calculus and Appell systems

I. Boson calculus

The boson calculus is based on the action of the operators of multiplication by x_i and differentiation, D_i, acting on polynomials. With $x = (x_1, \ldots, x_N)$, $D = (D_1, \ldots, D_N)$, define the basis $[\![n]\!] = x^n$, then, X_j denoting the operator of multiplication by x_j,

$$X_j[\![n]\!] = x_j x^n = x^{n+e_j}, \quad D_j[\![n]\!] = n_j x^{n-e_j}$$

And we have the commutator

$$D_j(X_j[\![n]\!]) - X_j(D_j[\![n]\!]) = [\![n]\!]$$

i.e., the commutation relations $[D_j, X_i] = \delta_{ij} I$ where I denotes the identity operator.

Now define the *boson operators* $\mathcal{R}_i$, $\mathcal{V}_i$ acting on the (abstract) basis $[\![n]\!]$ according to

$$\mathcal{R}_i[\![n]\!] = [\![n + e_i]\!], \quad \mathcal{V}_i[\![n]\!] = n_i[\![n - e_i]\!]$$

The $\mathcal{R}_i$ are *raising operators* and the $\mathcal{V}_i$, formal differentiation or *velocity operators*. They satisfy the commutation relations

$$[\mathcal{V}_i, \mathcal{R}_j] = \delta_{ij} I$$

The associative algebra generated by the $\{\mathcal{R}_i, \mathcal{V}_i\}$ is the *Heisenberg-Weyl algebra*. Notice that $\mathcal{V}_j[\![0]\!] = 0$ for all j. Applying the algebra to $[\![0]\!]$, the *vacuum state,* yields the *Fock space,* with basis $[\![n]\!]$.

We note the following, which is readily shown by induction

1.1 Proposition.   *The Heisenberg-Weyl algebra has basis* $[\![n\, m]\!] = \mathcal{R}^n \mathcal{V}^m$ *with*

$$\mathcal{R}_i[\![n\, m]\!] = [\![n + e_i\, m]\!], \quad \mathcal{V}_j[\![n\, m]\!] = n_j[\![n - e_j\, m]\!] + [\![n\, m + e_j]\!]$$

By the Jordan map, the operators $\mathcal{R}_i \mathcal{V}_j$ correspond to the elementary matrices having all zero entries except for a one at the ij position. The operator $\mathcal{R}_\mu \mathcal{V}_\mu$ corresponds to the identity matrix via the Jordan map. On the basis $[\![n]\!] = \mathcal{R}^n[\![0]\!]$ it acts according to

$$\mathcal{R}_\mu \mathcal{V}_\mu[\![n]\!] = |n|\, [\![n]\!]$$

with $|n| = \sum n_i$. Thus, this operator is called the *number operator* or the *degree operator* as it gives the total degree of the state.

Another family of boson operators, $\bar{\mathcal{R}}_i$, $\bar{\mathcal{V}}_i$, is defined by the action

$$\bar{\mathcal{R}}_i[\![n]\!] = (n_i + 1)[\![n + e_i]\!], \quad \bar{\mathcal{V}}_i[\![n]\!] = [\![n - e_i]\!]$$

These correspond to the operators multiplication by x and differentiation acting on the monomials $x^n/n!$.

II. Holomorphic canonical calculus

For $\phi(z) = \phi(z_1, \ldots, z_N)$, holomorphic in a neighborhood of $0 \in \mathbf{C}^N$, $\phi(z)$ is the *symbol* of the differential operator $\phi(D)$, acting on exponentials $e^{zx} = e^{z_\mu x_\mu}$ by

$$\phi(D)e^{zx} = \phi(z)e^{zx}$$

$\phi(D)$ acts on polynomials $p(x)$ by expanding in Taylor series around the origin so that

$$\phi(D)p(x) = \sum_{n \geq 0} \frac{p^{(n)}(x)\phi^{(n)}(0)}{n!}$$

with only a finite number of non-zero terms. Introducing X, the operator of multiplication by x, acting on polynomials we have the commutation relations

$$[\phi(D), X_j] = \frac{\partial \phi}{\partial D_j}, \quad [D_j, f(X)] = \frac{\partial f}{\partial X_j} \tag{2.1}$$

for polynomials f. We can extend these relations to the *Leibniz formula*

2.1 Proposition.   *For polynomials $f(X)$, we have the commutation rule:*

$$[\phi(D), f(X)] = \sum_{n \geq 0} \frac{f^{(n)}(X)\phi^{(n)}(D)}{n!}$$

Proof: This follows by induction on the degree of f using the relations (2.1).   ■

Now

2.2 Definition.   The algebra $\mathcal{E}$ consists of functions of the form (finite sums)

$$\sum p_k(x) e^{a_k x}$$

where p_k are polynomials and $a_k \in \mathbf{C}^N$.

Operators $\phi(D)$, X, and exponentials in X preserve $\mathcal{E}$. We have

$$\phi(D)p(x)e^{ax} = \sum_{n \geq 0} \frac{p^{(n)}(x)\phi^{(n)}(D)}{n!} e^{ax}$$

$$= \sum_{n \geq 0} \frac{p^{(n)}(x)\phi^{(n)}(a)}{n!} e^{ax}$$

Thus,

2.3 Proposition. *The Leibniz formula, Prop. 2.1, holds for $f \in \mathcal{E}$, as operators acting on the space $\mathcal{E}$.*

Unless ϕ is entire, the operator $\phi(D)$ acts only on exponentials e^{ax} such that a is in the domain of ϕ. In this sense, this operator calculus is local.

Extending this to locally holomorphic maps on domains in $\mathbf{C}^N$,

2.4 Definition. $V(z) = (V_1(z), \ldots, V_N(z))$ are *canonical coordinates* at 0 if V is holomorphic in a neighborhood of 0, with $V(0) = 0$, and the Jacobian matrix $V' = \left(\dfrac{\partial V_i}{\partial z_j} \right)$ nonsingular at 0. $U(v)$ denotes the functional inverse of V, i.e., $z_j = U_j(V(z))$. The *canonical dual* variables are given by

$$Y_j = X_\lambda W_{\lambda j}(D)$$

where $W(z) = V'(z)^{-1}$ is the matrix inverse to $V'(z)$.

2.5 Proposition. *The commutation relations*

$$[V_i(D), Y_j] = \delta_{ij} I$$

hold.

Proof: We have

$$[V_i(D), X_\lambda] W_{\lambda j} = (V')_{i\lambda} W_{\lambda j} = \delta_{ij} I$$

■

As the $V_i(D)$ mutually commute, to check that we indeed have a boson calculus, we need the commutativity of the Y's.

2.6 Proposition. *The variables $Y_1, \ldots, Y_N$ commute.*

Proof: Denoting differentiation by a comma followed by the appropriate subscript, we have

$$[Y_i, Y_j] = [X_\lambda W_{\lambda i}, X_\mu W_{\mu j}] = X_\lambda W_{\lambda i, \mu} W_{\mu j} - X_\mu W_{\mu j, \lambda} W_{\lambda i}$$
$$= X_\epsilon W_{\epsilon i, \mu} W_{\mu j} - X_\epsilon W_{\epsilon j, \mu} W_{\mu i}$$

I.e., we need to show that $W_{ki, \mu} W_{\mu j} = W_{kj, \mu} W_{\mu i}$, that the expression $W_{ki, \mu} W_{\mu j}$ is symmetric in ij. Recall that if W depends on a parameter, t, say, then $W = Z^{-1}$ satisfies $\dot{W} = -W \dot{Z} W$. Thus, the relation $W = (V')^{-1}$ yields the matrix equation

$$\frac{\partial W}{\partial z_i} = -W \frac{\partial V'}{\partial z_i} W$$

which gives

$$\frac{\partial W_{ki}}{\partial z_\mu} W_{\mu j} = -W_{k\varepsilon} \left(\frac{\partial V'}{\partial z_\mu}\right)_{\varepsilon\lambda} W_{\lambda i} W_{\mu j}$$

$$= -W_{k\varepsilon} \frac{\partial^2 V_\varepsilon}{\partial z_\mu \partial z_\lambda} W_{\lambda i} W_{\mu j}$$

so that the required symmetry follows from equality of the mixed partials of V.   ∎

The corresponding *canonical basis polynomials* are

$$\eta_n(x) = Y_1^{n_1} \cdots Y_N^{n_N} 1 \tag{2.2}$$

Thus

2.7 Proposition.   *The canonical holomorphic calculus is a boson calculus with the identifications $\mathcal{R}_i = Y_i$, $\mathcal{V}_i = V_i(D)$, $[\![0]\!] = 1$, and $[\![n]\!] = \eta_n(x)$. The number operator is given by YV.*

The basic fact is

2.8 Theorem.   *The following relations hold, for $v = (v_1, \dots, v_N)$,*

$$e^{vY} 1 = \sum_{n \geq 0} \frac{v^n}{n!} \eta_n(x) = e^{xU(v)}$$

Proof: The left-hand equality is the generating function version of eq. (2.2). To see the right-hand side, substitute $v = V(z)$ which gives the formulation $e^{zx} = \sum (V(z)^n/n!)\eta_n$. On the left side, the operator $V(D)$ acts as multiplication by the symbol $V(z)$. On the right, multiplication by $V(z)$ acts as the abstract boson operator $\mathcal{V}$, by shifting indices, while the holomorphic calculus says that $V(D)$ on η_n agrees with this action.   ∎

2.1   APPELL DUALITY

Using the canonical calculus we have the following duality that is an implicit form of Lagrange inversion.

2.1.1 Proposition.   *Given the canonical expansion $e^{xU(v)} = \sum (v^n/n!)\eta_n(x)$ and the corresponding expansion $e^{\alpha V(z)} = \sum (z^n/n!)\phi_n(\alpha)$, then:*

1. *If $\phi_n(\alpha) = \sum_m c_{nm}\alpha^m$, then $x^n = \sum_m c_{nm}\eta_m(x)$*

   and the inverse relations

2. *If $\eta_n(x) = \sum_m \eta_{nm}x^m$, then $\alpha^n = \sum_m \eta_{nm}\phi_m(\alpha)$*

Proof: Substitute the canonical variable Y for α in the expansion of $e^{\alpha V(z)}$ and apply to the vacuum 1 to find

$$e^{V(z)Y}\,1 = e^{xz} = \sum_{n\geq 0} \frac{z^n}{n!}\,\phi_n(Y)1$$

Writing

$$\phi_n(Y)1 = \sum_m c_{nm} Y^m 1 = \sum_m c_{nm}\eta_m(x)$$

yields the result.   ∎

Examples. 1. For the one-variable case a standard example is $V(z) = e^z - 1$. Then setting $e^{\alpha V(z)} = \sum_{n\geq 0} \frac{z^n}{n!}\,\phi_n(\alpha)$, we have

$$e^{xU(v)} = (1+v)^x = \sum_{n\geq 0} \frac{v^n}{n!}\,x(x-1)\cdots(x-n+1)$$

Thus, $x^n = \sum c_{nk} x(x-1)\cdots(x-k+1)$, with $\phi_n(\alpha) = \sum c_{nk}\alpha^k$. The c_{nk} are, of course, Stirling numbers. And expanding

$$e^{\alpha(e^z-1)} = e^{-\alpha}\sum_{m\geq 0}\frac{\alpha^m e^{mz}}{m!} = \sum_{n\geq 0}\frac{z^n}{n!}\,\mu_n(\alpha)$$

shows that we have the moment generating function of the Poisson distribution and that $\phi_n(\alpha) = \mu_n(\alpha)$, the moments.

2. In one variable, consider the Gaussian with drift, $V(z) = z - z^2/2$. Then $U(v) = \sqrt{1-2v}$. With the Hermite polynomials

$$H_n(x,\alpha) = \sum_k \binom{n}{2k}(-1)^k x^{n-2k}(2\alpha)^k(\tfrac{1}{2})_k$$

we have

$$e^{\alpha V(z)} = \sum_{n\geq 0}\frac{z^n}{n!}\,H_n(\alpha,\alpha)$$

i.e.,

$$\phi_n(\alpha) = \sum_k \binom{n}{2k}(-1)^k \alpha^{n-k} 2^k(\tfrac{1}{2})_k$$

Thus, rewriting in terms of factorials,

$$x^n = \sum_k \frac{n!(-1)^k}{(n-2k)!\,2^k k!}\,\eta_{n-k}(x)$$

where

$$e^{x\sqrt{1-2v}} = \sum_{n\geq 0} \frac{v^n}{n!}\, \eta_n(x)$$

These are essentially *Bessel polynomials*.

3. A multivariable version of Bessel polynomials thus can be constructed as follows. Fix $N \geq 1$, and take $V_j = z_j - \frac{1}{2}\sum z_i^2$, for $1 \leq j \leq N$. Squaring and summing, we find

$$z_j = V_j + S\,, \text{with } S = \frac{1 - \sum V_j - \sqrt{(1-\sum V_j)^2 - N\sum V_j^2}}{N}$$

We have the Jacobian

$$V' = I - Z$$

where Z is the rank-one matrix with every row equal to $z = (z_1,\ldots,z_N)$. Thus, the inverse can be readily found using geometric series and we have $W = I + (1-\sum z_j)^{-1}Z$. That is, the canonical variables are

$$Y_j = X_j + \left(\sum X_i\right)\frac{D_j}{1 - \sum D_i}$$

The corresponding canonical polynomials $\eta_n(x)$ are a family of multivariable Bessel polynomials and they are related to the Hermite polynomials in several variables as in Example 2.

2.2 θ-POLYNOMIALS

Observe that the canonical polynomials $\eta_n(x)$, for $N = 1$, being of the form $(xW)^n 1$ always have a common factor of x. Thus, one can define an associated sequence of θ-polynomials

$$\theta_n(x) = x^{-1}\,\eta_{n+1}(x)$$

In the situation of Example 2 above these give the standard version of the Bessel polynomials.

2.2.1 Proposition.   *The θ-polynomials satisfy:*

1. $\theta_n = W\eta_n$.

2. *The difference relation* $(1 - V')\theta_n = \theta_n - x\theta_{n-1}$.

3. *The generating function relation*

$$U'(v)\,e^{xU(v)} = \sum_{n\geq 0} \frac{v^n}{n!}\, \theta_n(x)$$

Proof: The first two properties follow from the definition. Property #3 follows by differentiating the generating function for the $\eta_n(x)$, $e^{xU(v)}$, and dividing out the factor x. ■

The operator $1 - V'$ appearing in the difference relation is natural with $V'(0) = 1$ as this gives a differential-difference relation for the θ-polynomials.

III. Canonical Appell systems

In one variable, Appell systems $\{h_n\}$ are typically defined by these properties:

$$h_n(x) \text{ is a polynomial of degree } n, \quad n \geq 0$$
$$Dh_n(x) = n\, h_{n-1}(x)$$

For $N \geq 1$, we have analogously

$$h_n(x) \text{ is a polynomial of degree } n, \quad n \geq 0$$
$$D_j h_n(x) = n_j\, h_{n-e_j}(x) \tag{3.1}$$

where degree n means that the polynomial has top term x^n and other terms are of lower (total) degree. The condition on the degree is a non-degeneracy assumption that will become clear below.

Let $\{h_n(x)\}$ be an Appell system. Let

$$F(z,x) = \sum_{n \geq 0} z^n\, h_n(x)/n!$$

be a generating function for this system. The basic property (3.1) implies

$$\frac{\partial F}{\partial x_i} = z_i\, F \tag{3.2}$$

In general we have the form
$$F(z,x) = e^{zx}\, G(z)$$

The expansion $G(z) = \sum_{n \geq 0} z^n\, c_n/n!$ yields

$$h_n(x) = \sum_{m \geq 0} \binom{n}{m} c_m x^{n-m}$$

as a generic expression for Appell polynomials (which may also be derived inductively from (3.1)). The condition on the degree gives us $c_0 \neq 0$, i.e., $G(0) \neq 0$.

Next we notice that (3.2) may be read from right to left, i.e., multiplication by z_i acts as differentiation D_i. Now consider the action of $\partial/\partial z_j$:

$$\frac{\partial F}{\partial z_j} = \sum_{n \geq 0} z^n \, h_{n+\mathbf{e}_j}(x)/n!$$

i.e., $\partial/\partial z_j$ acts as a raising operator: $h_n \to h_{n+\mathbf{e}_j}$. With $G(0) \neq 0$ we can locally express $G(z) = e^{H(z)}$ so that F takes the form

$$F(z,x) = e^{zx + H(z)}$$

where we normalize by $G(0) = 1$, $H(0) = 0$. The operators D_j and $\partial/\partial z_j$ satisfy

$$D_j \, F = z_j \, F, \qquad \frac{\partial F}{\partial z_j} = \left(x_j + \frac{\partial H}{\partial z_j} \right) F$$

Thus, X_j denoting the operator of multiplication by x_j,

$$h_{n+\mathbf{e}_j} = \left(X_j + \frac{\partial H}{\partial D_j} \right) h_n$$

In summary,

3.1 Theorem. *For Appell systems, given $H(z)$ an arbitrary function holomorphic in a neighborhood of 0, the boson calculus is given by $\mathcal{R}_i = X_i + \dfrac{\partial H}{\partial D_i}$, $\mathcal{V}_i = D_i$, with states $[\![n]\!] = h_n$. The h_n have the generating function*

$$e^{z\mathcal{R}} \, [\![0]\!] = e^{zx + H(z)} = \sum_{n \geq 0} \frac{z^n}{n!} \, h_n(x)$$

3.1 EVOLUTION EQUATION AND HAMILTONIAN FLOW

Now consider the evolution equation

$$\frac{\partial u}{\partial t} = H(D)u, \qquad u(x,0) = e^{zx}$$

with H locally holomorphic, as in the above discussion. We find

$$u(x,t) = e^{tH(D)} \, e^{zx} = e^{zx + tH(z)}$$

and expanding in powers of z, we have the Appell system

$$h_n(x,t) = e^{tH(D)} \, x^n$$

Note that in the previous section the t is absorbed into the H, alternatively, set to 1. The h_n satisfy $\partial u/\partial t = H(D)u$ with polynomial initial condition $u(x,0) = x^n$. Thus, we see Appell systems as *evolved powers*. The monomials x^n are built by successive multiplication by x_j, which we denote by the operators X_j: $X_j x^n = x^{n+e_j}$. Here we conjugate by the flow e^{tH}:

$$h_{n+e_j} = \left(e^{tH} X_j e^{-tH}\right) e^{tH} x^n = e^{tH} x^{n+e_j}$$

I.e., the raising operator is given by $\mathcal{R} = e^{tH} X e^{-tH}$. By the holomorphic operator calculus we have $[e^{tH}, X_j] = t(\partial H/\partial D_j) e^{tH}$, so that

$$\mathcal{R}_j = X_j + t\frac{\partial H}{\partial D_j}$$

as we have seen previously (for $t = 1$).

The mapping $(X, D) \to (\mathcal{R}, \mathcal{V})$ is given by the Heisenberg-Hamiltonian flow

$$\mathcal{R} = e^{tH} X e^{-tH}, \qquad \mathcal{V} = e^{tH} D e^{-tH}$$

which induces an automorphism of the entire Heisenberg-Weyl algebra. As t varies, writing $X(t)$ for $\mathcal{R}$, we have the *Heisenberg-Hamiltonian equations of motion* (suppressing subscripts)

$$\dot{X} = [H, X] = \frac{\partial H}{\partial D}, \qquad \dot{D} = [H, D] = -\frac{\partial H}{\partial X}$$

where in the case $H = H(D)$, D remains constant so that $\mathcal{V} = D$.

3.2 STOCHASTIC FORMULATION

Suppose that H comes from a family of probability measures p_t with corresponding random variables X_t by Fourier-Laplace transform as follows:

$$\langle e^{zX_t} \rangle = \int e^{zx} p_t(dx) = e^{tH(z)} \tag{3.2.1}$$

with $H(0) = 0$ here corresponding to the fact that the measures integrate to 1. Then

$$e^{zx+tH(z)} = \int e^{z(x+u)} p_t(du)$$

and

$$h_n(x,t) = \int (x+u)^n p_t(du) = \langle (x + X_t)^n \rangle \tag{3.2.2}$$

are *moment polynomials*.

3.2.1 Proposition.   *In the stochastic case,*

$$h_n(x,t) = \sum_{m \geq 0} \binom{n}{m} \mu_m(t) x^{n-m}$$

where $\mu_m(t)$ are moments of the probability measure p_t.

Proof: Expand out equation (3.2.2).   ∎

The probability measures satisfying eq. (3.2.1) form a convolution family: $p_t * p_s = p_{t+s}$, with the X_t a corresponding stochastic process. In this sense, we see the $h_n(x,t)$ as averages of the evolution of the functions x^n along the paths of the stochastic process X_t.

Remark. Unless the measures p_t are *infinitely divisible*, one will not be able to take t to be a continuous variable. But in any case, we *always* have Appell systems, analytic in t. What can be guaranteed is that if $e^{H(z)} = \int e^{zx} p(dx)$ then this extends to the discrete-parameter process for integer-valued $t \geq 0$. For other values of t, the corresponding measures will not necessarily be probability measures, i.e., positivity may not hold.

3.3 CANONICAL SYSTEMS

The principal feature of (X,D) in the construction of Appell systems is that they are boson variables. We can make Appell systems starting from any canonical pair (Y,V), $Y = XW$, and evolve under the Heisenberg-Hamiltonian flow

$$\dot{Y} = [H,Y], \qquad \dot{V} = [H,V]$$

For $H = H(D)$, $V = V(D)$ is invariant, while, writing $H' = (\partial H/\partial D_1, \ldots, \partial H/\partial D_N)$,

$$\begin{aligned} \mathcal{R} = Y(t) = e^{tH} X e^{-tH} W(D) &= (X + tH')W \\ &= Y + tH'W \end{aligned} \tag{3.3.1}$$

The canonical Appell system is thus $h_n(x,t) = e^{tH} \eta_n(x)$.

3.3.1 Theorem.   *For canonical Appell systems, we have:*

1. *The generating function*

$$e^{v\mathcal{R}} [\![0]\!] = e^{zU(v)+tH(U(v))} = \sum_{n \geq 0} \frac{v^n}{n!} h_n(x,t)$$

2. *The relation*

$$e^{V(z)\mathcal{R}} [\![0]\!] = e^{zz+tH(z)}$$

3. *The form of X*

$$X = \mathcal{R}V' - tH'$$

Proof: The first relation comes by applying $e^{tH(D)}$ to the generating function for the η_n, Theorem 2.8,

$$e^{tH}\,e^{zU(v)} = e^{zU(v)+tH(U(v))}$$

on the one hand, which is then the generating function for $h_n(x,t) = e^{tH}\,\eta_n(x)$. Relation #2 follows from #1 by replacing $v = V(z)$. For #3, recall eq. (3.3.1),

$$\mathcal{R} = (X + tH')W = (X + tH')(V')^{-1}$$

Solving for X yields the result.   ■

References

For more details and discussion, the reader is referred to Chapter 4 and Chapter 5, §I, of [19].

Chapter 2   Representations of Lie groups

With $\xi = (\xi_1, \ldots, \xi_d)$, $\{\xi_i\}$ a basis for the Lie algebra $\mathcal{G}$, a basis for the universal enveloping algebra $\mathcal{U}(\mathcal{G})$ is given by $[\![n]\!] = \xi^n = \xi_1^{n_1} \cdots \xi_d^{n_d}$ where the product is ordered, since the ξ_i do not commute in general. As we saw for Appell systems, it is natural to look at generating functions to see how multiplication by the basis elements ξ_i on $\mathcal{U}$ looks. We have

$$\sum_{n \geq 0} \frac{A^n}{n!} \xi^n = \sum \frac{(A_1 \xi_1)^{n_1}}{n_1!} \cdots \frac{(A_d \xi_d)^{n_d}}{n_d!} = e^{A_1 \xi_1} \cdots e^{A_d \xi_d} \tag{0.1}$$

This is an element of the group, as it is a product of the one-parameter subgroups generated by the basis elements.

For group elements near the identity we can use $A = (A_1, \ldots, A_d)$ as coordinates. Multiplication by an element ξ_j is realized as a vector field acting on functions of A. As this is dual to the action on the basis $[\![n]\!]$, we dualize again to find the action on the basis, which we express in terms of boson operators $\mathcal{R}, \mathcal{V}$.

I.   Coordinates on Lie groups

Write $X \in \mathcal{G}$ in the form $X = \alpha_\mu \xi_\mu$. Group elements in a neighborhood of the identity can be expressed as

$$e^X = e^{\alpha_\mu \xi_\mu} = g(A) = e^{A_1 \xi_1} e^{A_2 \xi_2} \cdots e^{A_d \xi_d} \tag{1.1}$$

The α_i are called *coordinates of the first kind* and the A_i *coordinates of the second kind.*

Now consider the one-parameter subgroup generated by X, e^{sX}. The coordinates α scale by the factor s, while the coordinates A can be thought of as functions of the single parameter s. With $X = \alpha_\mu \xi_\mu$, we have group elements

$$g(A(s)) = e^{sX} = e^{s\alpha_\mu \xi_\mu} \tag{1.2}$$

In particular, evaluating at $s = 1$ gives the coordinate transformation $A = A(\alpha)$ corresponding to (1.1). Differentiating (1.2) we have

$$\begin{aligned}
Xg &= \sum_i e^{A_1 \xi_1} \cdots e^{A_{i-1} \xi_{i-1}} \dot{A}_i \xi_i e^{A_i \xi_i} \cdots e^{A_d \xi_d} \\
&= \dot{A}_\mu \partial_\mu g
\end{aligned} \tag{1.3}$$

with $\partial_\mu = \partial/\partial A_\mu$, and differentiation with respect to s denoted by a dot. Observe that each element ξ_i appears in its proper place embedded in the ordered product.

Now we will see that the coordinates A contain the complete information about the Lie algebra structure.

1.1 Theorem.   *The coordinates $A(s)$ have the expansion*

$$A_i(s) = s\alpha_i - \frac{s^2}{2} \sum_{j,k} \alpha_j \alpha_k \theta_{jk} c^i_{jk} + \cdots$$

where θ denotes the ordering symbol.

Proof: $A(0) = 0$ by construction. Setting $s = 0$ in the relation $X e^{sX} = \dot{A}_\mu \partial_\mu g$, with $g = e^{sX}$, yields $X = \dot{A}_\mu(0)\xi_\mu$, i.e., $\dot{A}_i(0) = \alpha_i$. Taking the second derivative with respect to s yields

$$X^2 g = \ddot{A}_\mu \partial_\mu g + \dot{A}_\mu \dot{A}_\nu \partial_\mu \partial_\nu g$$

where the important feature is that when differentiating g, the ξ's remain in increasing order. Thus, evaluating at $s = 0$:

$$X^2 = \ddot{A}_\mu(0)\xi_\mu + \alpha_\lambda^2 \xi_\lambda^2 + 2\alpha_\lambda \alpha_\mu \theta_{\lambda\mu} \xi_\lambda \xi_\mu$$

In $X^2 = \alpha_\lambda \alpha_\mu \xi_\lambda \xi_\mu$, the terms where $\lambda > \mu$ yield $c^\nu_{\lambda\mu} \xi_\nu \theta_{\mu\lambda}$. Interchanging λ and μ in this last expression picks up a minus sign. Cancelling common terms yields the result.   ∎

Note that if we are given A as functions of α, then we find $A(s)$ immediately by scaling $\alpha \to s\alpha$.

The *group law* is written in terms of the coordinates A as

$$g(A)g(A') = g(A \odot A')$$

The group elements $g(A(s))$ in equation (1.2) form an abelian subgroup. This gives the *addition formula*

$$g(A(s) \odot A(s')) = g(A(s + s')) \tag{1.4}$$

II. Dual representations

Multiplication by basis elements ξ, $g \to g\xi$, are realized as left-invariant vector fields ξ^*, acting on functions of A. (Left-invariant means that the action commutes with multiplication by group elements on the left, thus the ξ acts on the right.) They are given in terms of the *pi-matrix* $\pi^*(A)$, by $\xi^*_i = \pi^*_{i\mu}(A)\partial_\mu$. The left-invariant vector field corresponding to X is $X^* = \alpha_\mu \xi^*_\mu$. Similarly, multiplying on the left gives right-invariant vector fields, $\xi^\dagger_i = \pi^\dagger_{i\mu}(A)\partial_\mu$.

2.1 Definition.   The *dual representations* are realizations of the Lie algebra as vector fields in terms of the coordinates of the second kind acting on the left or right respectively.

The vector fields are given in terms of the pi-matrices $\pi^\dagger(A), \pi^*(A)$ according to:

$$\xi_j\, g(A) = \pi^\dagger_{j\mu}(A)\partial_\mu g(A)\,, \qquad g(A)\xi_j = \pi^*_{j\mu}(A)\partial_\mu g(A) \tag{2.1}$$

Now we have our main tool,

2.2 Lemma. *Splitting Lemma*

Denote by $\pi(A)$ either pi-matrix, π^* or $\pi^{\dagger}$. Then we have

$$\dot{A}_k = \alpha_{\lambda}\pi_{\lambda k}(A)$$

with initial values $A_k(0) = 0$.

Proof: First, we have, acting by X on the left,

$$\dot{g} = Xg = \alpha_{\mu}\xi_{\mu}\,g = \alpha_{\mu}\xi_{\mu}^{\dagger}g = \alpha_{\lambda}\pi_{\lambda\mu}^{\dagger}\partial_{\mu}g$$

And from the right,

$$\dot{g} = gX = g\,\alpha_{\mu}\xi_{\mu} = \alpha_{\mu}\xi_{\mu}^{*}g = \alpha_{\lambda}\pi_{\lambda\mu}^{*}\partial_{\mu}g$$

In general, if A depends on a parameter s, then for any function $f(A)$, we have the flow $\dot{f} = \dot{A}_{\mu}\partial_{\mu}f$. Thus $\dot{g} = \dot{A}_{\mu}\partial_{\mu}g$, cf. eq. (1.3), and the result follows. ∎

We have the basic fact

2.3 Proposition.   The pi-matrices satisfy $\pi(0) = I$, the identity.

Proof: We have seen that $\dot{A}(0) = \alpha$. Setting $s = 0$ in Lemma 2.2 yields the result. ∎

The right dual mapping $\xi \to \xi^*$ gives a Lie homomorphism, i.e., $[\xi_i, \xi_j]^* = [\xi_i^*, \xi_j^*]$, while the action on the left reverses the order of operations, giving a Lie antihomomorphism $[\xi_i, \xi_j]^{\dagger} = [\xi_j^{\dagger}, \xi_i^{\dagger}]$. An important feature is that the left and right actions commute. Thus, as vector fields, *every $\xi^{\dagger}$ commutes with every ξ^*.*

We have for the Lie bracket of these vector fields:

Right dual   For $[\xi_i^*, \xi_j^*]$, evaluating at $A = 0$ and using the above Proposition:

$$\pi_{i\mu}^{*}\partial_{\mu}\pi_{jk}^{*} - \pi_{j\mu}^{*}\partial_{\mu}\pi_{ik}^{*} = c_{ij}^{\mu}\pi_{\mu k}^{*}$$
$$\partial_i\pi_{jk}^{*}(0) - \partial_j\pi_{ik}^{*}(0) = c_{ij}^{k} \tag{2.2}$$

Left dual   Similarly, for $[\xi_i^{\dagger}, \xi_j^{\dagger}]$:

$$\pi_{i\mu}^{\dagger}\partial_{\mu}\pi_{jk}^{\dagger} - \pi_{j\mu}^{\dagger}\partial_{\mu}\pi_{ik}^{\dagger} = c_{ji}^{\mu}\pi_{\mu k}^{\dagger}$$
$$\partial_i\pi_{jk}^{\dagger}(0) - \partial_j\pi_{ik}^{\dagger}(0) = c_{ji}^{k}$$

As well, we see that by construction, the left and right actions commute, so that we have

Combined   Commuting the left and right yields

$$\pi^*_{i\mu}\partial_\mu\pi^\dagger_{jk} = \pi^\dagger_{j\mu}\partial_\mu\pi^*_{ik}$$

$$\partial_i\pi^*_{jk}(0) = \partial_j\pi^\dagger_{ik}(0)$$

Combining this last relation with (2.2), we have

$$\partial_i\pi^*_{jk}(0) - \partial_i\pi^\dagger_{jk}(0) = c^k_{ij}$$

the transpose of the adjoint representation. This can be summarized as follows

2.4 Proposition.   *The transposed adjoint representation is the linearization of the difference between the right and left duals.*

Thus

2.5 Definition.   The *extended adjoint representation* is the difference of the right and left duals:

$$\tilde{\xi}_j = \xi^*_j - \xi^\dagger_j$$

This gives a representation of the Lie algebra, since $\xi_j \to -\xi^\dagger_j$ gives a Lie homomorphism, the minus sign reversing the commutators. And we set the corresponding $\tilde{\pi}$:

$$\tilde{\pi}(A) = \pi^*(A) - \pi^\dagger(A)$$

2.1 AFFINE TYPE ALGEBRAS

There are some general features of the dual representations that can be remarked depending on the structure of the Lie algebra. Here we look at the case where there is an abelian ideal. In the next section we look at chains of subalgebras.

2.1.1 Definition.   If there is an abelian ideal, which we denote by $\mathcal{P}$, then we say that the Lie algebra is of *general affine-type.*

The basis elements for $\mathcal{P}$ are $R = (R_1, \ldots, R_N)$. The algebra has the form $\mathcal{G} = \mathcal{P} \oplus \mathcal{K}$, with corresponding group factorization $\mathbf{G} = \mathbf{PK}$. Since $\mathcal{K}$ normalizes $\mathcal{P}$, we have the adjoint action $\mathrm{ad}_{\mathcal{P}}K$ for $K \in \mathcal{K}$, which gives a representation of $\mathcal{K}$ on $\mathcal{P}$.

2.1.2 Proposition.   For $X = R_0 + K$, with $R_0 \in \mathcal{P}$, $K \in \mathcal{K}$,

$$e^{sX} = e^{V(s)R} e^{sK}$$

where $V(s)R$ is given by

$$V(s)R = \int_0^s e^{u\,\mathrm{ad}_{\mathcal{P}}K} \, du \, R_0$$

Proof: Writing $e^{sX} = e^{V(s)R} e^{A(s)K}$, we have

$$\dot{g} = g(R_0 + K) = \dot{V}Rg + g\dot{A}K$$
$$= \left(e^{A(s)\text{ad}\,K} R_0\right) g + gK$$

This gives $\dot{A} = 1$, so that $A(s) = s$, while $\dot{V}R = e^{A(s)\text{ad}\,K} R_0$ and hence the result.

∎

Note that this does not depend upon, nor gives any information about, the splitting for the subgroup $\mathbf{K}$.

2.2 LIE FLAGS

In a vector space, $\mathcal{W}$, a *flag* is an increasing sequence of subspaces $\mathcal{W}_i$ such that each is of codimension one in the next:

$$\mathcal{F} = \{\, \mathcal{W}_0, \mathcal{W}_1, \ldots, \mathcal{W}_d \,\}$$

$\mathcal{W}_0 = \{\, 0 \,\}$, $\mathcal{W}_d = \mathcal{W}$, with

$$\mathcal{W}_0 \subset \mathcal{W}_1 \subset \mathcal{W}_2 \subset \cdots \subset \mathcal{W}_d$$

A flag corresponds to an *adapted basis* $\{\, \xi_1, \xi_2, \ldots, \xi_d \,\}$ such that

$$\mathcal{W}_k = \text{span}\,\{\, \xi_1, \xi_2, \ldots, \xi_k \,\}$$

We define an *increasing Lie flag* to be a flag of *subalgebras*

$$\mathcal{B}_0 \subset \mathcal{B}_1 \subset \mathcal{B}_2 \subset \cdots \subset \mathcal{B}_d$$

i.e., each $\mathcal{B}_j$ is closed under the Lie bracket as well. With respect to the corresponding adapted basis, matrices of the adjoint representation have zeros in a lower left block. Given a basis $\{\, \xi_1, \xi_2, \ldots, \xi_d \,\}$, consider the corresponding flag and define

$$\mathcal{B}_j^k = \text{span}\,\{\, \xi_j, \xi_{j+1}, \ldots, \xi_k \,\}$$

with $\mathcal{B}_k = \mathcal{B}_1^k$. A flag such that $\mathcal{B}_j^d$ are subalgebras for $j = d - 1, d - 2, \ldots, 1$ is a *decreasing Lie flag*. A flag that is both increasing and decreasing is a *symmetric flag*. From $\mathcal{B}_j^k = \mathcal{B}_1^k \cap \mathcal{B}_j^d$ follows

2.2.1 Proposition. *A symmetric flag has the property that for all $1 \leq j < k \leq d$, $\mathcal{B}_j^k$ is a subalgebra.*

(Throughout, the term flag will be understood to refer to a Lie flag.)

Note that by reversing the order of the basis an increasing flag corresponds to a decreasing flag and vice versa.

Every solvable, in particular every nilpotent, algebra has an increasing Lie flag consisting of ideals $\mathcal{B}_i$.

2.2.1 Dual representations and flags

Given an increasing flag, with adapted basis $\{\,\xi_1,\ldots,\xi_d\,\}$, denote by $\check{\xi}_j^*$ the transpose of the matrix of ξ_j in the adjoint representation *restricted to the subalgebra* $\mathcal{B}_j$. I.e., columns $j+1$ through d of $\check{\xi}_j$ are zero'd out and then the matrix is transposed. Dually, for a decreasing flag, we denote by $\check{\xi}_j^{\dagger}$ the transposed matrix of the restriction of the adjoint action of ξ_j to the subalgebra $\mathcal{B}_j^d$, i.e., the first j columns are zero'd out, then the matrix transposed.

2.2.1.1 Theorem. *For the dual representations we have:*

(i) *Given an increasing flag, the pi-matrix for the right dual is given by*

$$\pi^*(A) = \exp(A_d\check{\xi}_d^*)\exp(A_{d-1}\check{\xi}_{d-1}^*)\cdots\exp(A_1\check{\xi}_1^*)$$

(ii) *Given a decreasing flag, the pi-matrix for the left dual is given by*

$$\pi^{\dagger}(A) = \exp(-A_1\check{\xi}_1^{\dagger})\exp(-A_2\check{\xi}_2^{\dagger})\cdots\exp(-A_d\check{\xi}_d^{\dagger})$$

Proof: First we note that the proof of (ii) is dual to that of (i) in the sense of left-right. The proof of (i) is by induction. Denoting

$$E_{d-1} = \exp(A_1\xi_1)\cdots\exp(A_{d-1}\xi_{d-1})$$

with g as usual the product up to d, we have

$$
\begin{aligned}
g\xi_k &= E_{d-1}\left(e^{A_d\operatorname{ad}\xi_d}\,\xi_k\right)e^{A_d\xi_d}\\
&= E_{d-1}\exp(A_d\check{\xi}_d)_{\lambda k}\xi_\lambda e^{A_d\xi_d}\\
&= \exp(A_d\check{\xi}_d)_{\lambda k}\left(\pi^*_{\lambda\mu}(d-1|d-1)\partial_\mu E_{d-1}\right)e^{A_d\xi_d} + g\exp(A_d\check{\xi}_d)_{dk}\xi_d
\end{aligned}
$$

which should equal

$$\left(\exp(A_d\check{\xi}_d^*)\pi^*(d-1|d)\right)_{k\mu}\partial_\mu g$$

where $\pi^*(d-1|d-1)$ is the right dual pi-matrix for the action of $\{\,\xi_1,\ldots,\xi_{d-1}\,\}$ on the subalgebra $\mathcal{B}_{d-1}$ and $\pi^*(d-1|d)$ denotes its extension to $\mathcal{B}_d$. By construction (as part of induction hypothesis) the dth row of $\pi^*(d-1|d)$ is zero except for a dd entry of 1. Since $\mathcal{B}_{d-1}$ is a subalgebra, the first $d-1$ elements of the dth row of $\check{\xi}_j$ vanish if $j<d$, which translates into the vanishing of the corresponding column entries of the transpose. Thus, for the kd term:

$$
\begin{aligned}
\left(\exp(A_d\check{\xi}_d^*)\pi^*(d-1|d)\right)_{kd}\partial_d g &= \exp(A_d\check{\xi}_d)_{\lambda k}\pi^*(d-1|d)_{\lambda d}g\xi_d\\
&= \exp(A_d\check{\xi}_d)_{dk}g\xi_d
\end{aligned}
$$

as required. Thus the result. ∎

2.3 DIFFERENTIAL RELATIONS

We will show that the right dual vector fields generate the action of the group on functions. We will explicitly indicate the dependence of g on ξ by writing $g(A, \xi)$ for $g(A)$. Then we can indicate by writing ξ^* for ξ that we are using the particular realization of the Lie algebra given by the right dual representation.

2.3.1 Lemma. Denote by $g(A', \xi^*)$ the group element $g(A')$ with ξ replaced by ξ^*. Then for smooth functions f,

$$g(A', \xi^*) f(A) = f(A \odot A')$$

Proof: We will use the expansion $g(A, \xi) = 1 + A_\mu \xi_\mu + \cdots$. We can write $g(A'(s), \xi^*) = e^{s \alpha'_\mu \xi^*_\mu}$, so that we are in fact considering an expression of the form $e^{sX^*} f(A)$, then evaluating at $s = 1$. Thus, by the method of characteristics, it is sufficient to consider the case $f(A) = A$. Then we have

$$g(A', \xi^*) g(A, \xi) = g(A, \xi) g(A', \xi) = g(A \odot A', \xi)$$

and thus

$$g(A', \xi^*) \left[1 + A_\mu \xi_\mu + \cdots \right] = 1 + (g(A', \xi^*) A_\mu) \xi_\mu + \cdots$$
$$= 1 + (A \odot A')_\mu \xi_\mu + \cdots$$

which yields $g(A', \xi^*) A_i = (A \odot A')_i$ as required. ∎

And we have the following differential relations connecting the left and right duals.

2.3.2 Theorem. Let $u = f(A \odot A')$ for a smooth function f. Then u satisfies the differential equations

$$\xi^\ddagger(A') u = \xi^*(A) u$$

where the variable acted upon is indicated explicitly.

Proof: By Lemma 2.3.1, we have $u = g(A', \xi^*(A)) f(A)$. So we look at the action of $\xi^\ddagger(A')$ on $g(A', \xi^*(A))$. By definition of the left dual,

$$\xi^\ddagger(A') g(A', \xi^*(A)) = \xi^\ddagger(A') g(A', \xi) \big|_{\xi = \xi^*(A)} = \xi \, g(A', \xi) \big|_{\xi = \xi^*(A)}$$
$$= \xi^*(A) g(A', \xi^*(A))$$

as required. ∎

2.4 INVARIANTS

The relations $gX = Xg$ and the splitting lemma, 2.2, yield the identities

$$\alpha_\mu \pi^*_{\mu j}(A(s)) = \alpha_\mu \pi^\dagger_{\mu j}(A(s))$$

which hold for all $s \geq 0$. These yield relations between the A's and a's that are invariant, i.e., are independent of s.

2.5 GROUP ACTION AND SOLUTIONS

The *group-theoretical method of solution* of a differential equation (or system) is based on solving a relatively simple system and then applying the action of a group. For example, if $H\phi = \psi$, then $(gHg^{-1})(g\phi g^{-1}) = g\psi g^{-1}$. With regard to the splitting formula, we are interested in solving the differential system $\dot{A} = \alpha\pi(A)$. If we have, e.g., $\xi_1, \ldots, \xi_N$ mutually commuting, then with $X = \sum_1^N \alpha_i \xi_i$, we know that $A_i = \alpha_i$ for $1 \leq i \leq N$. With knowledge of the group law, or adjoint action, say, by direct matrix multiplication, we can compute $e^{s\tilde{X}}$, with $\tilde{X} = \tilde{g}X\tilde{g}^{-1}$, using the adjoint action of the group element $\tilde{g}$. We will apply this technique later in some specific cases.

Example. As a standard technique, to solve $\dot{x} = Hx$, with H a real symmetric matrix, the system separates by using the action of the orthogonal group to diagonalize H.

2.6 DOUBLE DUAL

Acting on the group elements considered as generating functions for the basis of $\mathcal{U}(\mathcal{G})$, eq. (0.1), we have multiplication by A_j acting (dually) on the basis as the velocity operator $\mathcal{V}_j$, while differentiation ∂_j dualizes to the raising operator $\mathcal{R}_j$. Now we can write the vector fields of the dual representations in terms of boson operators $\mathcal{R}, \mathcal{V}$.

The vector field $\pi(A)\partial$ becomes $\mathcal{R}\pi(\mathcal{V})$. Thus, the left and right *double duals*

$$\hat{\xi}_j = \mathcal{R}_\mu \pi^\dagger_{j\mu}(\mathcal{V}), \quad \hat{\xi}^*_j = \mathcal{R}_\mu \pi^*_{j\mu}(\mathcal{V})$$

where we drop the dagger for the left double dual and just call it the 'double dual.' For the left action, the double dual thus gives a Lie homomorphism. These give the left and right multiplication by the basis elements ξ_j on the enveloping algebra in terms of the basis $[\![n]\!]$. I.e.,

$$\xi_j[\![n]\!] = \hat{\xi}_j[\![n]\!], \qquad [\![n]\!]\xi_j = \hat{\xi}^*_j[\![n]\!]$$

The pi-matrices for the double duals are effectively the transposes of the original pi-matrices.

2.6.1 Definition. We denote the transposed pi-matrices by

$$\hat{\pi} = (\pi^\dagger)^\dagger, \qquad \hat{\pi}^* = (\pi^*)^\dagger$$

so that

2.6.2 Theorem. *The double duals give the left and right multiplication by the basis elements ξ_j on the enveloping algebra in terms of the basis $[\![n]\!]$. I.e.,*

$$\xi_j[\![n]\!] = \hat{\xi}_j[\![n]\!], \qquad [\![n]\!]\xi_j = \hat{\xi}_j^*[\![n]\!]$$

with

$$\hat{\xi}_j = \mathcal{R}_\mu \hat{\pi}_{\mu j}(\mathcal{V}), \qquad \hat{\xi}_j^* = \mathcal{R}_\mu \hat{\pi}_{\mu j}^*(\mathcal{V})$$

2.7 ADJOINT GROUP

We can calculate the exponential of the adjoint representation in terms of the π matrices. First we remark that the exponential of the adjoint representation connects the right and left duals:

$$g\xi_j = g\xi_j g^{-1} g = \xi_j^* g = Ad_g(\xi_j)g = Ad_g(\xi_j^\dagger)g \qquad (2.7.1)$$

where Ad_g denotes the exponential of the adjoint representation, conjugation by g. It is given explicitly as the matrix $g(A, \check{\xi})$ (acting on the row vector formed from the basis vectors). Next we have the following Proposition which is immediate from the definitions, equation (2.1).

2.7.1 Proposition. *The left and right duals are related by*

$$\xi^* = \pi^* \pi^{\dagger^{-1}} \xi^\dagger$$

where ξ denotes the column array with entries ξ_i.

2.7.2 Definition. Define the matrix

$$\check{\pi} = g(A, \check{\xi})$$

the exponential of the adjoint representation.

Now for the main result of this section.

2.7.3 Theorem.   *The exponential of the adjoint representation, $g(A, \check{\xi})$, is given by*

$$\tilde{\pi} = \hat{\pi}^{-1} \hat{\pi}^*$$

Proof: From equation (2.7.1), we have the matrix of Ad_g, acting on the basis, given by the transpose of $g(A, \check{\xi})$. I.e.,

$$\xi^* = \tilde{\pi}^\dagger \xi^\ddagger \tag{2.7.2}$$

Comparing with Proposition 2.7.1, we have $\tilde{\pi}^\dagger = \pi^* \pi^{\dagger^{-1}}$. And the result follows by definition of the matrices $\hat{\pi}$, $\hat{\pi}^*$ as transposes, Definition 2.6.1.   ∎

2.8   GENERAL HAMILTONIANS

We can use the double duals to find 'Hamilton's equations' for the Lie algebra. Let $H(\xi)$ be a function of ξ, e.g., an element of $\mathcal{U}(\mathcal{G})$ or a combination of exponentials, with coefficients in $\mathcal{U}(\mathcal{G})$, cf., the algebra $\mathcal{E}$, Ch. 1, Definition 2.2. We want to solve

$$\frac{\partial u}{\partial t} = [H, u]$$

with initial condition $u(0) = \xi_j$. We have

$$\dot{\xi}_j = [H, \xi_j] = H\xi_j - \xi_j H$$
$$= (\hat{\xi}_j^* - \hat{\xi}_j)H$$
$$= \mathcal{R}_\mu \tilde{\pi}_{j\mu}(\mathcal{V})H$$

where on the right-hand sides, everything is evaluated at $\xi = \xi(t)$.

2.9   COADJOINT ORBITS

For calculating the coadjoint orbits, or effectively what is the same, to calculate the exponential of the adjoint representation, the matrices π are sufficient. Namely, use the exponential of the adjoint representation, Theorem 2.7.3.

III.   Matrix elements

Matrix elements for the representation of the group acting on the enveloping algebra correspond to $g(A)$ acting on the monomial basis $[\![n]\!] = \xi_1^{n_1} \cdots \xi_d^{n_d}$. Recall equation (0.1). Define the matrix elements $\left\langle \begin{smallmatrix} m \\ n \end{smallmatrix} \right\rangle_A$ according to

$$g(A)[\![n]\!] = \sum_m \left\langle \begin{smallmatrix} m \\ n \end{smallmatrix} \right\rangle_A [\![m]\!] \tag{3.1}$$

Our main result is that the matrix elements can be computed recursively.

3.1 PRINCIPAL FORMULA

Using the right dual, we have, with $(\xi^*)^n = (\xi_1^*)^{n_1} \cdots (\xi_d^*)^{n_d}$,

3.1.1 Theorem. *Principal formula for computing the matrix elements:*

$$\left\langle \begin{matrix} m \\ n \end{matrix} \right\rangle_A = (\xi^*)^n c_m(A) \tag{3.1.1}$$

Proof: We use induction on the index n. For $n = 0$, we have $g[\![0]\!] = \sum c_m(A)[\![m]\!]$, i.e., $\left\langle \begin{matrix} m \\ 0 \end{matrix} \right\rangle = c_m(A)$. Now we observe that as functions of the index n, both $\left\langle \begin{matrix} m \\ n \end{matrix} \right\rangle$ and $(\xi^*)^n c_m(A)$ satisfy the same recurrence relations. For the left-hand side of (3.1.1), apply ξ_j^* to both sides of relation (3.1). We have

$$\xi_j^* g[\![n]\!] = g\xi_j[\![n]\!] = g\hat{\xi}_j[\![n]\!] = \sum_m R_\mu \hat{\pi}_{\mu j}(\mathcal{V}) \left\langle \begin{matrix} m \\ n \end{matrix} \right\rangle_A [\![m]\!]$$

on the one hand, with only the index n being affected, and on the other,

$$\xi_j^* g[\![n]\!] = \xi_j^* \sum_m \left\langle \begin{matrix} m \\ n \end{matrix} \right\rangle_A [\![m]\!] = \sum_m \xi_j^* \left\langle \begin{matrix} m \\ n \end{matrix} \right\rangle_A [\![m]\!]$$

where here only the A-variables are affected. For the right-hand side of (3.1.1), since the right dual is a Lie homomorphism, we have directly from Theorem 2.6.2 that

$$\xi_j^* (\xi^*)^n = R_\mu \hat{\pi}_{\mu j}(\mathcal{V})(\xi^*)^n$$

with only the index n coming into play. ∎

The group law $g(A)g(A') = g(A \odot A')$ in terms of the matrix elements reads

$$\left\langle \begin{matrix} m \\ \lambda \end{matrix} \right\rangle_A \left\langle \begin{matrix} \lambda \\ n \end{matrix} \right\rangle_{A'} = \left\langle \begin{matrix} m \\ n \end{matrix} \right\rangle_{A \odot A'} \tag{3.1.2}$$

i.e., this is a matrix representation of the group. And on the coordinates, c_m, we have the action

$$\sum_n \left\langle \begin{matrix} m \\ n \end{matrix} \right\rangle_A c_n(A') = c_m(A \odot A') \tag{3.1.3}$$

This can be seen directly by multiplying $c_n(A')$ by $g(A)$ on the left and by $[\![n]\!]$ on the right and summing:

$$g(A) \sum_n c_n(A')[\![n]\!] = \sum_{m,n} c_n(A') \left\langle \begin{matrix} m \\ n \end{matrix} \right\rangle_A [\![m]\!]$$

while the left-hand side equals $g(A)g(A') = g(A \odot A') = \sum_m c_m(A \odot A')[\![m]\!]$ as required. Equation (3.1.3) can be interpreted as saying that $c_m(A \odot A')$ is the generating function for the matrix elements $\left\langle \begin{matrix} m \\ n \end{matrix} \right\rangle_A$.

As a Corollary of Theorem 3.1.1, we have

3.1.2 Proposition.   *The right dual pi-matrices are matrix elements for transitions between basis elements. I.e.,*

$$\pi_{ij}^* = \left\langle \begin{matrix} \mathbf{e}_j \\ \mathbf{e}_i \end{matrix} \right\rangle$$

Proof: This follows from the principal formula thus

$$\left\langle \begin{matrix} \mathbf{e}_j \\ \mathbf{e}_i \end{matrix} \right\rangle = \xi_i^* A_j = \pi_{i\lambda}^* \partial_\lambda A_j = \pi_{ij}^*$$

∎

3.2 DIFFERENTIAL RECURRENCES

Using the double duals, we can find differential recurrence relations in which the action of the vector fields is expressed in terms of a corresponding action on the discrete indices of the matrix elements. To emphasize the action on the indices, we will use boldface to indicate which indices the operator is acting on. First, define the action of the ξ by duality on the index m of the matrix elements $\left\langle \begin{matrix} m \\ n \end{matrix} \right\rangle$, denoting this by $\bar{\xi}$, thus:

$$\sum_m \left\langle \begin{matrix} m \\ n \end{matrix} \right\rangle \xi_j [\![m]\!] = \sum_m \bar{\xi}_j \left\langle \begin{matrix} m \\ n \end{matrix} \right\rangle [\![m]\!] \tag{3.2.1}$$

3.2.1 Theorem.   *For the left and right dual representations we have the differential recurrence relations:*

$$\xi_j^* \left\langle \begin{matrix} m \\ n \end{matrix} \right\rangle = \hat{\xi}_j \left\langle \begin{matrix} m \\ \mathbf{n} \end{matrix} \right\rangle$$
$$\xi_j^\dagger \left\langle \begin{matrix} m \\ n \end{matrix} \right\rangle = \bar{\xi}_j \left\langle \begin{matrix} \mathbf{m} \\ n \end{matrix} \right\rangle$$

with the A-variables understood.

Proof: The proof of the right dual action has been given above in the proof of Theorem 3.1.1. For the left action, we have

$$\xi_j g[\![n]\!] = \xi_j^\dagger \sum_m \left\langle \begin{matrix} m \\ n \end{matrix} \right\rangle [\![m]\!] = \sum_m \left\langle \begin{matrix} m \\ n \end{matrix} \right\rangle \xi_j [\![m]\!]$$

which comes back to equation (3.2.1).   ∎

Recall the boson operators $\bar{\mathcal{R}}, \bar{\mathcal{V}}$ corresponding to the action of X and D on $x^n/n!$, equivalently, to multiplication by A and differentiation, ∂, on $c_n(A)$.

3.2.2 Corollary. *The action of $\bar{\xi}$ on $\left\langle\begin{smallmatrix} m \\ n \end{smallmatrix}\right\rangle_A$ is the same as that of $\xi^{\ddagger}$ on $c_m(A)$, i.e.,*

$$\bar{\xi}\left\langle\begin{matrix} m \\ n \end{matrix}\right\rangle_A = \pi^{\ddagger}(\bar{\mathcal{R}})\bar{\mathcal{V}}\left\langle\begin{matrix} \mathbf{m} \\ n \end{matrix}\right\rangle_A$$

Proof: From the principal formula, Theorem 3.1.1, $\left\langle\begin{smallmatrix} m \\ n \end{smallmatrix}\right\rangle_A = (\xi^*)^n c_m(A)$, we have directly $\xi_j^{\ddagger}\left\langle\begin{smallmatrix} m \\ n \end{smallmatrix}\right\rangle_A = (\xi^*)^n \xi_j^{\ddagger} c_m(A)$, since the left and right duals commute. ∎

IV. Induced representations and homogeneous spaces

Homogeneous spaces are quotient spaces, such as cosets of the group modulo a subgroup. Here we have a subalgebra, $\mathcal{Q}$, say, and take a quotient of $\mathcal{U}(\mathcal{G})$ mapping the subalgebra to zero or to scalars. This induces a representation of $\mathcal{G}$, on the reduced basis, and a representation of **G** as well, giving a homogeneous space with coordinates corresponding to the complement of $\mathcal{Q}$ in $\mathcal{G}$.

4.1 REPRESENTATIONS FROM THE DOUBLE DUAL

In terms of the double dual, the basis for $\mathcal{U}(\mathcal{G})$ is given by $\mathcal{R}^n[\![0]\!]$, i.e., the double dual acts on polynomials in $\mathcal{R}$. As these are commuting variables, we map them $\mathcal{R}_i \to y_i$ and $\mathcal{V} \to \partial/\partial y_i$, so that the double dual acts directly on polynomials in the variables $\{y_1,\ldots,y_d\}$. Corresponding to taking quotients of the universal enveloping algebra by ideals, one forms *induced representations* via the double dual by setting some variables equal to scalars, as parameters, and some variables equal to zero. In particular, irreducible representations may be constructed.

Example. The Heisenberg algebra has basis $\{\xi_1, \xi_2, \xi_3\}$, with commutation relations $[\xi_3, \xi_1] = \xi_2$, with ξ_2 central. The double dual has the form (see Ch. 6 for details) $\hat{\xi}_1 = \mathcal{R}_1$, $\hat{\xi}_2 = \mathcal{R}_2$, $\hat{\xi}_3 = \mathcal{R}_3 + \mathcal{R}_2\mathcal{V}_1$. With the mapping described above, we have the double dual acting on functions of y:

$$\hat{\xi}_1 = y_1\,, \quad \hat{\xi}_2 = y_2\,, \quad \hat{\xi}_3 = y_3 + y_2\partial_1$$

Note, e.g., that the span of $\{y_1^{n_1} y_2^{n_2} y_3^{n_3}\}$ with $n_2 \geq p$, $n_3 \geq q$ for fixed $p, q \geq 0$ are invariant subspaces for the action of the Lie algebra. Writing y for y_1, taking $y_2 \to c$, a scalar, and $y_3 \to 0$, we have the representation on polynomials in the single variable y

$$\xi_1 f(y) = y f(y), \quad \xi_2 f(y) = c f(y), \quad \xi_3 f(y) = c f'(y)$$

which is an irreducible representation.

We state three principles:

1. Any variable not appearing explicitly in $\pi^{\ddagger}$ maps to a scalar.

2. Any variable appearing in $\pi^{\ddagger}$ that is mapped to a scalar forces the coefficients of the corresponding partial derivative in the double dual to be zero.

3. If two basis elements map to scalars acting on Ω, then their commutator must map to zero on Ω.

The first point follows from the fact that in the double dual no derivatives corresponding to 'missing' variables will appear. The second point is that if a variable acts as a scalar parameter, then it cannot have an explicit corresponding differentiation. The third point is seen by observing that if $\xi_i \Omega = \tau_i \Omega$, $i = 1, 2$, for scalars τ_i, then

$$[\xi_1, \xi_2]\Omega = \xi_1 \tau_2 \Omega - \xi_2 \tau_1 \Omega = 0$$

Note that from the first point follows that we can always map elements of the center to scalars, since none of their corresponding variables will appear in $\pi^{\ddagger}$, as long as a basis for the center forms part of the basis chosen for the algebra.

Generally, we choose an abelian subalgebra with basis $\{R_1, \ldots, R_N\}$, with dual coordinates V_j (i.e., denoted above by A_j, $1 \leq j \leq N$). We have group elements of the form

$$e^X = e^{V_1 R_1} \cdots e^{V_N R_N} e^{A_{N+1} \xi_{N+1}} \cdots e^{A_d \xi_d}$$

which gives via the double dual

$$e^{\dot{X}} = e^{V_1 y_1} \cdots e^{V_N y_N} e^{A_{N+1} \hat{\xi}_{N+1}} \cdots e^{A_d \hat{\xi}_d}$$

To get a representation of the algebra acting on functions of $(y_1, \ldots, y_N)$ we want the remaining elements of the basis, $\hat{\xi}_{N+1}$ to $\hat{\xi}_d$ to evaluate to scalars acting on a vacuum state, according to the principles noted above.

4.2 QUOTIENT REPRESENTATIONS

We consider quotients of the universal enveloping algebra. These correspond to factorization of the Lie group into chosen subgroups and then taking representations induced from representations of some of the subgroups.

The Lie algebra $\mathcal{G}$ is decomposed as a direct sum of three subalgebras: $\mathcal{P}$, $\mathcal{K}$, and $\mathcal{L}$. We denote the bases for these as follows: $\mathcal{P} = \text{span}\{R_i\}$, $\mathcal{K} = \text{span}\{\rho_i, \rho_{ij}\}$, and $\mathcal{L} = \text{span}\{L_i\}$. We decompose $\mathcal{K}$ further according to $\mathcal{H} = \text{span}\{\rho_i\}$ and $\mathcal{N} = \text{span}\{\rho_{ij}\}$. The corresponding dual variables (the A's, and α's in general) are denoted: V_i (v_i), H_i (h_i), C_{ij} (κ_{ij}), B_i (β_i). I.e., we have for a typical element $X \in \mathcal{G}$,

$$X = \alpha_\mu \xi_\mu = v_\lambda R_\lambda + h_\mu \rho_\mu + \kappa_{\lambda\mu} \rho_{\lambda\mu} + \beta_\nu L_\nu$$

and the corresponding factorization of the group $\mathbf{G} = \mathbf{PKL} = \mathbf{PHNL}$, with group elements

$$e^X = e^{V_\lambda R_\lambda} e^{H_\mu \rho_\mu} \dots e^{C_{ij} \rho_{ij}} \dots e^{B_\lambda L_\lambda}$$

In general, we will denote group coordinates of $\mathbf{K}$ by A, when the refinement into $\mathbf{K} = \mathbf{HN}$ is not necessary to be made explicit. The subalgebra $\mathcal{L}$ may be empty, but if it is not empty, it is in 1-1 correspondence with $\mathcal{P}$ and satisfies $[\mathcal{L}, \mathcal{P}] \subset \mathcal{K}$. The subalgebras $\mathcal{P}$, $\mathcal{L}$, and $\mathcal{H}$ are abelian, and we denote $N = \dim \mathcal{P}$, $r + r' = \dim \mathcal{K}$, with $r = \dim \mathcal{H}$, $r' = \dim \mathcal{N}$. It is required further that $\mathcal{P} \oplus \mathcal{K}$ and $\mathcal{L} \oplus \mathcal{K}$ be subalgebras. Note that in the double dual we have $\hat{R}_i = \mathcal{R}_i$. The R's are *raising operators* and the L's are *lowering operators*.

We take an abstract vector Ω, as $[\![0]\!]$, the *vacuum state*, or *cyclic vector*, and the representation space has the form $\mathcal{U}(\mathcal{G})\Omega$. An induced representation is formed by the prescription

$$L_j\Omega = 0, \quad \rho_i\Omega = \tau_i\Omega, \quad \rho_{ij}\Omega = 0$$

for all $1 \leq j \leq N$, $1 \leq i \leq r$, $1 \leq i,j \leq r'$, where τ_i are scalars, the *weights* of the representation. The action of the group elements now has the form

$$g(V, A, B)\Omega = e^X \Omega = e^{\tau_\mu H_\mu} e^{V_\lambda R_\lambda} \Omega \tag{4.2.1}$$

Remark. For simplicity, we write $e^{\tau H}$ for $e^{\tau_\mu H_\mu}$, with a similar interpretation for e^{VR}.

4.2.1 Matrix elements for quotient representations

We denote the matrix elements of the action of the group on the basis $R^n\Omega$ by $\left\langle\!\!\left\langle \begin{smallmatrix} m \\ n \end{smallmatrix} \right\rangle\!\!\right\rangle$. As in the principal formula, Theorem 3.1.1, we can use the right dual.

4.2.1.1 Proposition.   *For the quotient representation, the matrix elements are given by*

$$\left\langle\!\!\left\langle \begin{matrix} m \\ n \end{matrix} \right\rangle\!\!\right\rangle = (R^*)^n e^{\tau H} c_m(V)$$

Proof: To see this, pull the R's across the group action as in the principal formula:

$$gR^n\Omega = (R^*)^n g\Omega = (R^*)^n e^{\tau_\mu H_\mu} e^{V_\lambda R_\lambda} \Omega$$

as in equation (4.2.1). Now expand e^{VR} and pull out the coefficient of $R^m\Omega$ to find the result.   ∎

The differential recurrences, Theorem 3.2.1, simplify somewhat.

4.2.1.2 Proposition.   *The matrix elements $\left\langle\!\!\left\langle \begin{smallmatrix} m \\ n \end{smallmatrix} \right\rangle\!\!\right\rangle$ satisfy*

$$R^* \left\langle\!\!\left\langle \begin{matrix} m \\ n \end{matrix} \right\rangle\!\!\right\rangle = \mathcal{R} \left\langle\!\!\left\langle \begin{matrix} m \\ \mathbf{n} \end{matrix} \right\rangle\!\!\right\rangle, \qquad R^\ddagger \left\langle\!\!\left\langle \begin{matrix} m \\ n \end{matrix} \right\rangle\!\!\right\rangle = \bar{V} \left\langle\!\!\left\langle \begin{matrix} m \\ n \end{matrix} \right\rangle\!\!\right\rangle$$

Proof: The first relation follows from Proposition 4.2.1.1, the second from the fact that $R_j^\dagger = \partial/\partial V_j$. ∎

For finding the matrix elements for the quotient representation we have the following general observation.

4.2.1.3 Proposition. *Consider the product of group elements,*

$$g(V, A, B)g(V', 0, 0) = g(V'', A'', B'')$$

Then $e^{\tau H''} c_m(V'')$ is a generating function for the matrix elements $\left\langle\!\left\langle \begin{matrix} m \\ n \end{matrix} \right\rangle\!\right\rangle$. I.e.,

$$e^{\tau H''} c_m(V'') = \sum_n c_n(V') \left\langle\!\left\langle \begin{matrix} m \\ n \end{matrix} \right\rangle\!\right\rangle$$

Proof: From Proposition 4.2.1.1, we have, by Lemma 2.3.1,

$$\sum_n c_n(V') \left\langle\!\left\langle \begin{matrix} m \\ n \end{matrix} \right\rangle\!\right\rangle = e^{V'R^*} e^{\tau H} c_m(V) = f((V, A, B) \odot (V', 0, 0))$$

where $f(V, A, B) = e^{\tau H} c_m(V)$. Thus the result. ∎

The representation property $g_1 g_2 \Omega = (g_1 g_2)\Omega$ yields the following relations

4.2.1.4 Proposition. *For the product of group elements*

$$g(V, H, C, B)g(V', H', C', B') = g(V''', H''', C''', B''')$$

we have

$$H''' = H' + H'', \qquad V''' = V''$$

Proof: Apply $g(V, H, C, B)$ to both sides of the relation

$$g(V', H', C', B')\Omega = e^{\tau H'} e^{V'R} \Omega = e^{\tau H'} g(V', 0, 0, 0)\Omega$$

Then on the left side the action on Ω of the composition of the group elements yields the result. ∎

References

The basic reference on enveloping algebras is Dixmier [10]. B. Gruber has used the method of quotients of the enveloping algebra to obtain representations, e.g., see Gruber & Klimyk [31].

The idea of factorizing the group elements is found in Wei & Norman [53], [54].

Duchamp & Krob [11],[12] discuss the structure of the enveloping algebra of a free Lie algebra.

The recursive calculation of the matrix elements of the group was formulated in [24].

Chapter 3   General Appell systems

I.   Convolution and stochastic processes

Given two measures on $\mathbf{R}^N$, p_1, p_2, the convolution $p_1 * p_2$ is defined by the relation

$$\int f(x)\,(p_1 * p_2)(dx) = \iint f(x + x')\,p_1(dx)p_2(dx')$$

for all bounded continuous functions f. In fact it is sufficient that this relation hold for all complex exponentials e^{iax}, with $a \in \mathbf{R}^N$.

NOTE: Throughout our discussion, we assume that probability measures are in the class $\mathcal{P}^*$, where $p \in \mathcal{P}^*$ means that the Fourier-Laplace transform $\phi(z) = \int e^{zx}\,p(dx)$, $z \in \mathbf{C}^N$, is analytic in a neighborhood of the origin in $\mathbf{C}^N$.

WARNING: In the discussion below, we will use X and X_i to denote random variables, not general elements of a Lie algebra $\mathcal{G}$ as in Ch. 2.

For measures in $\mathcal{P}^*$, we have, for $z \in \mathbf{C}^N$,

$$\int e^{zx}\,(p_1 * p_2)(dx) = \phi_1(z)\phi_2(z) \tag{1.1}$$

where $\phi_i(z) = \int e^{zx}\,p_i(dx)$, $i = 1, 2$. Start with p, $\phi(z) = \int e^{zx}\,p(dx)$. Since $\phi(0) = 1$, locally we can express $\phi(z) = e^{H(z)}$, with H locally analytic. Now form the convolution family p_t inductively by $p_{t+1} = p_t * p$, for integer t. If p is infinitely divisible, this may be extended to continuous $t \geq 0$. In any case, we set p_0 equal to the delta function at the origin. From eq. (1.1) we have

$$e^{tH(z)} = \int e^{zx}\,p_t(dx)$$

Corresponding to the measures p_t of a convolution family, we have random variables $X(t)$. In discrete time, $X(t)$ is the sum of t independent, identically distributed random variables, each with distribution p_1. In the continuous-time case, the random variables $X(t)$ form a process with stationary independent increments, i.e., the distribution of $X(t) - X(s)$ depends only on the difference $t - s$, and the increments $X(t) - X(s), X(t') - X(s')$ are independent if the intervals $(s, t), (s', t')$ are disjoint.

Given a measure p, with Fourier-Laplace transform $\phi(z)$, we have the convolution operator

$$\phi(D)\,f(x) = \int f(x + x')p(dx')$$

For the convolution family p_t we have

$$e^{tH(D)} f(x) = \int f(x + x') p_t(dx)$$

To extend this to Lie groups, define the convolution operator

$$\int f(A \odot A') p(dA')$$

and the convolution of measures, $p_1 * p_2$, by

$$\int f(A) (p_1 * p_2)(dA) = \iint f(A \odot A') p_1(dA) p_2(dA')$$

To express this in terms of operators, we think of $\phi(D)$ as the Fourier-Laplace transform of the group element $e^{x'D}$ and use the action $e^{x'D} f(x) = f(x + x')$ as follows:

$$\int f(x + x') p(dx') = \int e^{x'D} p(dx') f(x)$$
$$= \phi(D) f(x)$$

For Lie groups, we have, recalling Ch. 2, Lemma 2.3.1,

$$\int f(A \odot A') p(dA') = \int g(A', \xi^*) f(A) p(dA') \tag{1.2}$$
$$= \phi(\xi^*) f(A)$$

Now we will see one of the features of using coordinates of the second kind. The measure p has an associated random vector $X \in \mathbf{R}^d$, $X = (X_1, \ldots, X_d)$. If the components X_i are independent, then the measure p factors into one-dimensional marginal distributions, $p^{(i)}$, corresponding to the X_i. Since

$$g(A', \xi^*) = e^{A_1' \xi_1^*} \cdots e^{A_d' \xi_d^*}$$

we can integrate out each variable A_i' separately. We thus have

1.1 Proposition. *If the components X_i are independent with distributions $p^{(i)}$, $1 \le i \le d$, and $\phi_i(z_i) = e^{H_i(z_i)}$ are the corresponding Fourier-Laplace transforms, then:*

1. *The operator relation*

$$\langle g(X, \xi) \rangle = e^{H_1(\xi_1)} \cdots e^{H_d(\xi_d)}$$

   where the angle brackets denote integration with respect to the measure p.

2. *The convolution operator formulation*

$$\int f(A \odot A') p(dA') = e^{H_1(\xi_1^*)} \cdots e^{H_d(\xi_d^*)} f(A)$$

This holds, at least, for functions $f \in \mathcal{E}$, the class defined in Ch. 1, Definition 2.2.

Suppose we have $\langle g(X) \rangle = e^{H_1(\xi_1)} \cdots e^{H_d(\xi_d)}$. The difference in the non-commutative case is that if you have independent X, X' then multiplying

$$\langle g(X)g(X') \rangle = e^{H_1(\xi_1)} \cdots e^{H_d(\xi_d)} e^{H_1'(\xi_1)} \cdots e^{H_d'(\xi_d)}$$

on the right hand side one cannot just re-order the terms and combine H_i with H_i'. It does follow, however, that if $\langle g(X) \rangle = \phi(\xi)$, then if $X(t)$ corresponds to the t-th convolution power of p, for integer t, then

$$\langle g(X(t)) \rangle = (\phi(\xi))^t \tag{1.3}$$

Remark. Even though we cannot generally rewrite $e^A e^B$ as e^{A+B}, the Trotter product formula indicates how to write e^{A+B} using the semigroups generated by e^A and e^B. Namely, $(e^{(1/n)A} e^{(1/n)B})^n \to e^{A+B}$ as $n \to \infty$. This extends as well to more than two operators, as will be useful below. The indicated result holds for bounded operators, and has various extensions to unbounded operators.

II.   Stochastic processes on Lie groups

First we want to see the form of the left-hand side of eq. (1.3).

2.1 Definition.   A *signal* means either 1) a deterministic trajectory in $\mathbf{R}^d$, in discrete or continuous time, or 2) the trajectory of a stochastic process in $\mathbf{R}^d$ with stationary independent increments, in discrete or continuous time, having independent components. In all cases, we take the origin as starting point.

We will map a signal to the Lie group by the exponential map using the signal as coordinates of the second kind. Take a signal $\alpha(t) = (\alpha_1(t), \ldots, \alpha_d(t))$. For discrete time, we write

$$\alpha(t) = \sum_{0 < k \le t} \Delta\alpha(k)$$

with $\Delta\alpha(k) = \alpha(k) - \alpha(k-1)$. In continuous time we have $\alpha(t) = \int_0^t d\alpha$, where for α of bounded variation, integrals with respect to $d\alpha$ are Stieltjes integrals, and for stochastic processes, they are stochastic integrals. For discretization of the continuous time process, fix $n > 0$, and set

$$\Delta\alpha(k) = \int_{t_{k-1}^{(n)}}^{t_k^{(n)}} d\alpha = \alpha(t_k^{(n)}) - \alpha(t_{k-1}^{(n)})$$

where $t_k^{(n)} = \frac{kt}{n}$ is an equi-partition of the interval $[0,t]$ with step-size n^{-1}. Thus, $\alpha(t) = \sum_{0 < k \le n} \Delta\alpha(k)$.

Mapping into the Lie group, define the multiplicative increment

$$g(\Delta\alpha(k)) = e^{\Delta\alpha_1(k)\xi_1} \ldots e^{\Delta\alpha_d(k)\xi_d}$$

For the discrete time process, consider the product over the time interval $[0,t]$:

$$g(A(t)) = \prod_k g(\Delta\alpha(k)) \tag{2.1}$$

For the continuous time case, we take the limit as $n \to \infty$ of these expressions, yielding the product integral:

$$g(A(t)) = \prod g(d\alpha)$$

This construction shows how the Lie structure and the signal interact. Thinking of the Lie group as a 'black box' into which we feed the signal, we say that $A(t)$ is the *Lie response* to the input $\alpha(t)$.

2.2 Theorem. *For any α, the Lie response satisfies:*

1. *For discrete time processes*

$$A(t) = \sum_{0 \le k < t} \left[A(k) \odot \Delta\alpha(k+1) - A(k) \right]$$

2. *For continuous time processes*

$$A(t) = \int_0^t \left(A(s) \odot d\alpha - A(s) \right)$$

in the sense that the corresponding group element exists if and only if the integral exists.

Proof: Write dt meaning an increment of one for discrete-time processes, or indicating that the limit of increments $\Delta t \to 0$ is taken ($n \to \infty$ in the setup of the previous paragraph), with a similar interpretation of $d\alpha$. Then

$$g(A(t+dt)) = g(A(t))g(d\alpha) = g(A(t) \odot d\alpha)$$

so that $A(t+dt) = A(t) \odot d\alpha$. Subtracting $A(t)$ and integrating, i.e., summing and letting $\Delta t \to 0$, yields the result.   ∎

And specifically if we have a smooth signal

2.3 Theorem. *For smooth $\alpha(t)$,*

$$dA = d\alpha\, \pi^*(A)$$

and the Lie response has the form

$$A(t) = \int_0^t d\alpha(s)\, \pi^*(A(s))$$

Proof: By Lemma 2.3.1 of Ch. 2, for $A = A(s)$, the jth component

$$(A \odot d\alpha)_j = g(d\alpha, \xi^*) A_j$$
$$= A_j + d\alpha_\mu \xi_\mu^* A_j + \cdots = A_j + d\alpha_\mu \pi_{\mu j}^* + \cdots$$

where for a smooth signal the higher order terms vanish in the limit $\Delta t \to 0$. ∎

Now for the stochastic case, write $\alpha(t) = \tilde{X}(t)$, where the $\mathbf{R}^d$-valued process $\tilde{X}(t)$ satisfies

$$\langle e^{z \tilde{X}(t)} \rangle = e^{t H_1(z_1)} \cdots e^{t H_d(z_d)} = e^{t \sum H_i(z_i)}$$

With $\tilde{X}^{(n)}(t)$ denoting the level n discretization, we have, for each k, by time-homogeneity,

$$\langle e^{z \Delta \tilde{X}^{(n)}(k)} \rangle = e^{\frac{t}{n} \sum H_i(z_i)} \tag{2.2}$$

2.4 Theorem. *Let $\tilde{X}(t)$ be a continuous time stochastic signal in $\mathbf{R}^d$ such that $\langle e^{z \tilde{X}(t)} \rangle = e^{t \sum H_i(z_i)}$. The Lie response satisfies $X(t) = \int_0^t (X(s) \odot d\tilde{X}(s) - X(s))$. The group-valued process has expected value*

$$\langle g(X(t)) \rangle = e^{t \sum H_i(\xi_i)}$$

Proof: The equations for the Lie response are given according to Theorem 2.2. For the group element, recalling Proposition 1.1, eqs. (2.1) and (2.2) give, using time-homogeneity:

$$\langle g(X^{(n)}(t)) \rangle = \langle g(\Delta X(1)) \rangle^n$$
$$= \left(e^{\frac{t}{n} H_1(\xi_1)} \cdots e^{\frac{t}{n} H_d(\xi_d)} \right)^n$$

Taking limits, the left-hand side yields the average of the Lie response, while on the right-hand side the limit converges as required via the Trotter product formula. ∎

And as in eq. (1.2),

2.5 Corollary. *For functions $f \in \mathcal{E}$, we have*

$$\langle f(A \odot X(t)) \rangle = e^{t H(\xi^*)} f(A)$$

In particular, for $f(A) = c_n(A)$, this yields the correlation functions for the Lie response process $X(t)$.

We use various realizations of $\mathcal{G}$ whether as differential operators or matrices and apply the relation of Theorem 2.4 acting on the appropriate vector space.

2.1 WIENER AND POISSON PROCESSES

The basic processes with independent increments are the *Wiener process,* often called *Brownian motion,* and the *Poisson process.* We denote the Wiener process by $w(t) = (w_1(t), \ldots, w_d(t))$ and the Poisson process by $\nu(t) = (\nu_1(t), \ldots, \nu_d(t))$. Each component process is a one-dimensional Wiener/Poisson process respectively and they are mutually independent. The stochastic differentials dw are *white noise* processes. The differential $d\nu_i$ of a Poisson process is a delta function at the jump times of the process, with independent exponentially distributed waiting times between the jumps.

Wiener process: The measures $p_t(dx) = p_t(x)\,dx$ are given by densities on $\mathbf{R}^d$

$$p_t(x) = e^{-|x|^2/(2t)} / (2\pi t)^{d/2}$$

with $H(z) = \frac{1}{2}\sum z_j^2$. Thus, Theorem 2.4 reads

$$\langle g(W(t)) \rangle = \exp\left(\frac{t}{2}\sum \xi_j^2\right)$$

where $W(t)$ is the Lie response to the white noise input. This shows one generalization of the Laplacian $\sum D_j^2$ of Euclidean space.

Poisson process: The marginal distributions of the components are given by

$$p_t^{(i)}(dx) = e^{-t}\sum_{k=0}^{\infty}\frac{t^k}{k!}\delta_k(dx)$$

with δ_k indicating the point mass at k. We have $H(z) = e^z - 1$, and for the Lie response

$$\langle g(N(t)) \rangle = \exp\left(t\sum(e^{\xi_i} - 1)\right)$$

Each component ν_i jumps randomly, independently of the other processes, and only one of them jumps at a time at which point the process on the group is moved by the corresponding factor e^{ξ_i}. This gives a random walk on the group, with random jump times.

III. Appell systems on Lie groups

For Lie groups, we would like to define Appell systems, say as in eq. (3.2.2) of Ch. 1, using convolution operators on powers of x. The problem is that it is not clear what to use as the analog of x^n. However, going back to the generating function, we have

$$e^{zx + tH(z)} = e^{tH(D)}e^{zx} = \langle e^{XD}e^{zx} \rangle$$

the expected value of the product of group elements. Using the coordinates A here, corresponding to the variable z in the above equation, we consider

$$\langle g(X)g(A)\rangle, \qquad \langle g(A)g(X)\rangle$$

where we have now a 'left system' and a 'right system', depending on which side the random variables are situated. Now,

$$g(X)g(A) = g(X \odot A) = \sum c_n(X \odot A)[\![n]\!]$$

The expected value of this relation should be the generating function for the Appell system. Here we have a non-commutative generating function and we see that the analog of x^n is given by the coefficients $c_n(A)$, which after all are $A^n/n!$.

3.1 Definition. The left and right Appell systems are, respectively,

$$h_n^L(A) = \langle c_n(X \odot A)\rangle, \quad h_n^R(A) = \langle c_n(A \odot X)\rangle$$

In the abelian case, the composition is addition and we recover the original Appell systems of Ch. 1. By definition these satisfy

$$\langle g(X)\,g(A)\rangle = \sum h_n^L(A)[\![n]\!]$$
$$\langle g(A)\,g(X)\rangle = \sum h_n^R(A)[\![n]\!]$$

In the first equation, with g acting on the left, the factored form requires reversing the order of terms if we want to express this in terms of the left dual $\xi^\dagger$. For the second equation, from Lemma 2.3.1 of Ch. 2, we have

3.2 Proposition. *The right Appell system may be expressed as*

$$h_n^R(A) = \langle g(X,\xi^*)\rangle c_n(A)$$

Recalling Proposition 1.1, we have

3.3 Proposition. *For $\langle g(X)\rangle = e^{H_1(\xi_1)} \cdots e^{H_d(\xi_d)} \equiv E(H(\xi))$ we have*

$$E(H(\xi))\,g(A) = \sum h_n^L(A)\,[\![n]\!]$$
$$g(A)\,E(H(\xi)) = \sum h_n^R(A)\,[\![n]\!]$$

And, here indicating explicitly time-dependence of the h_n,

3.4 Proposition. *For the case of processes, as in Theorem 2.4, with $\langle g(X(t))\rangle = e^{tH(\xi)}$,*

$$e^{tH(\xi)}\,g(A) = \sum h_n^L(A,t)\,[\![n]\!]$$
$$g(A)\,e^{tH(\xi)} = \sum h_n^R(A,t)\,[\![n]\!]$$

And we see that

3.5 Proposition. *The right Appell system solves* $u_t = H(\xi^*)u$, *for initial conditions* $c_n(A)$. *The left Appell system solves* $u_t = H(\xi^\dagger)u$, *for initial conditions* $c_n(A)$.

Proof: For the right system, the action of $H(\xi)$ on the right becomes $H(\xi^*)$. Since $H(\xi) = \sum H_i(\xi_i)$, acting on the left we can replace ξ by $\xi^\dagger$ since products involving distinct $\xi_i^\dagger$ will appear symmetrically so that the reversal of order has no effect. ∎

Now recall equation (3.1.3) of Ch. 2 (rewritten here with m, n interchanged):

$$\sum_m \left\langle \begin{matrix} n \\ m \end{matrix} \right\rangle_A c_m(A') = c_n(A \odot A')$$

This shows that

3.6 Proposition. *The left and right Appell systems satisfy:*

$$h_n^L(A) = \sum_m c_m(A) \langle \left\langle \begin{matrix} n \\ m \end{matrix} \right\rangle_X \rangle$$

$$h_n^R(A) = \sum_m \left\langle \begin{matrix} n \\ m \end{matrix} \right\rangle_A \langle c_m(X) \rangle$$

Comparing this with the standard Appell polynomials with the moments of the underlying probability distribution coming in, we see that in fact the rôle of the powers x^n is played by the matrix elements $\left\langle \begin{matrix} n \\ m \end{matrix} \right\rangle$ as well as c_n. This suggests defining *matrix Appell systems*, which we now consider.

3.1 MATRIX APPELL SYSTEMS

Similar to Definition 3.1, we have

3.1.1 Definition. *Matrix Appell systems*

$$^L H_n^m(A) = \langle \left\langle \begin{matrix} m \\ n \end{matrix} \right\rangle_{X \odot A} \rangle, \quad ^R H_n^m(A) = \langle \left\langle \begin{matrix} m \\ n \end{matrix} \right\rangle_{A \odot X} \rangle$$

From the fact that we have a matrix representation of the group, eq. (3.1.2) of Ch. 2:

3.1.2 Proposition.

$$^L H_n^m(A) = \langle \left\langle \begin{matrix} m \\ \lambda \end{matrix} \right\rangle_X \rangle \left\langle \begin{matrix} \lambda \\ n \end{matrix} \right\rangle_A$$

$$^R H_n^m(A) = \left\langle \begin{matrix} m \\ \lambda \end{matrix} \right\rangle_A \langle \left\langle \begin{matrix} \lambda \\ n \end{matrix} \right\rangle_X \rangle$$

The differential recurrences, Ch. 2, Theorem 3.2.1, combined with the above Proposition yield

3.1.3 Proposition.   *Matrix Appell systems satisfy the differential recurrence relations:*

$$\xi^* \, {}^L H_n^m(A) = \hat{\xi} \, {}^L H_n^m(A)$$

$$\xi^\ddagger \, {}^R H_n^m(A) = \bar{\xi} \, {}^R H_n^m(A)$$

3.2   TRANSFORMATION PROPERTIES

The transformation properties of the c_n and the matrix elements $\left\langle {}^m_n \right\rangle$ lead to interesting properties of the Appell systems. Using associativity of the group law, we can derive transformation properties of these Appell systems. Considering transformations of the c_n involving three variables, we find

3.2.1 Theorem.   *Appell systems h^L and h^R have the following transformation properties:*

$$h_n^R(A \odot A') = \left\langle {n \atop \lambda} \right\rangle_A h_\lambda^R(A')$$

$$\left\langle {n \atop \lambda} \right\rangle_A h_\lambda^L(A') = {}^R H_\lambda^n(A) c_\lambda(A')$$

$$h_n^L(A \odot A') = {}^L H_\lambda^n(A) c_\lambda(A')$$

Proof: For the first equation, start with $\langle c_n(A \odot A' \odot X) \rangle$. Then,

$$\langle c_n((A \odot A') \odot X) \rangle = h_n^R(A \odot A')$$

while

$$\langle c_n(A \odot (A' \odot X)) \rangle = \langle \left\langle {n \atop \lambda} \right\rangle_A c_\lambda(A' \odot X) \rangle$$

The second equation follows similarly from $\langle c_n(A \odot X \odot A') \rangle$. In one case the expected value is taken of the matrix elements, giving matrix Appell elements, in the other case, it gives the h^L. And similarly for the third equation from $\langle c_n(X \odot A \odot A') \rangle$.   ∎

Observe that the right Appell system h^R transforms according to the representation given by the matrix elements the same as the c_n. That is, they transform as a *vector* under the representation. Now we see that the matrix Appell systems transform covariantly, i.e. they form *tensor operators*.

3.2.2 Theorem.   *Matrix Appell systems ${}^L H$ and ${}^R H$ transform according to*

$${}^L H_n^m(A \odot A') = {}^L H_\lambda^m(A) \left\langle {\lambda \atop n} \right\rangle_{A'}$$

$${}^R H_\lambda^m(A) \left\langle {\lambda \atop n} \right\rangle_{A'} = \left\langle {m \atop \lambda} \right\rangle_A {}^L H_n^\lambda(A')$$

$${}^R H_n^m(A \odot A') = \left\langle {m \atop \lambda} \right\rangle_A {}^R H_n^\lambda(A')$$

Proof:

$$
{}^L H_n^m(A \odot A') = \langle\!\langle {m \atop n} \rangle\!\rangle_{X \odot A \odot A'}
$$

$$
= \langle\!\langle {m \atop \lambda} \rangle\!\rangle_{X \odot A} \langle {\lambda \atop n} \rangle_{A'}
$$

$$
= {}^L H_\lambda^m(A) \langle {\lambda \atop n} \rangle_{A'}
$$

And similarly via $\langle\!\langle {m \atop n} \rangle\!\rangle_{A \odot X \odot A'}$ and $\langle\!\langle {m \atop n} \rangle\!\rangle_{A \odot A' \odot X}$ for the remaining equations. ∎

Recalling Propositions 3.2, 3.3 and 3.4, we have a boson calculus for the right system.

3.2.3 Proposition.   *Let $\langle g(X, \xi^*) \rangle = E(\xi^*)$ be of the form $e^{H(\xi^*)}$ or a product of such factors. Then, the boson calculus for the right system is given by $[\![n]\!] = h_n^R$, with raising and velocity operators*

$$
\bar{\mathcal{R}}_j = E(\xi^*) A_j E(\xi^*)^{-1}
$$

$$
\bar{\mathcal{V}}_j = E(\xi^*) \partial_j E(\xi^*)^{-1}
$$

Proof: The operators $\bar{\mathcal{R}}, \bar{\mathcal{V}}$ appear since they are conjugate to the action of multiplication by A_j and ∂_j on $c_n(A) = A^n/n!$. ∎

References

A general reference for probability on groups is Heyer [34]. See Bougerol [4], [5] for studies of limiting behavior of probability measures on groups.

The works Fliess [27] and Fliess & Normand-Cyrot [28] show the use of non-commuting variables. See Gill [29] and Gill & Johansen [30] for product integration and applications.

This chapter is based on the articles [25], [26], [22].

Chapter 4   Canonical systems in several variables

In this chapter we focus on Lie algebras of *symmetric type*, described in the first section. From these algebras, the quotient representations yield homogeneous spaces, in particular symmetric spaces. The generating functions for the basis of the associated representations are called *coherent states*. There are connections with the multinomial distribution in probability theory and associated orthogonal polynomials generalizing the Meixner classes to several variables. We conclude with a Fourier technique for constructing new classes of orthogonal systems from certain given systems.

I.   Homogeneous spaces and Cartan decompositions

A symmetric Lie algebra has a Cartan decomposition $\mathcal{G} = \mathcal{P} \oplus \mathcal{K} \oplus \mathcal{L}$ into subalgebras $\mathcal{P}, \mathcal{K}, \mathcal{L}$ with the following properties:

$$[\mathcal{L}, \mathcal{P}] \subset \mathcal{K}, \quad [\mathcal{K}, \mathcal{P}] \subset \mathcal{P}, \quad [\mathcal{K}, \mathcal{L}] \subset \mathcal{L}$$

That is, $\mathcal{L}$ maps $\mathcal{P}$ into $\mathcal{K}$ by the adjoint map while $\mathcal{K}$ normalizes each of $\mathcal{P}$ and $\mathcal{L}$. As in Ch. 2, §4.2, we have $\mathcal{K} = \mathcal{H} \oplus \mathcal{N}$. (For semisimple algebras, $\mathcal{H}$ is the Cartan subalgebra.) We assume that $\mathcal{P}, \mathcal{L}$, and $\mathcal{H}$ are abelian. $\mathcal{P}$ and $\mathcal{L}$ are in one-one correspondence, with common dimension N. We have $\dim \mathcal{H} = r$, $\dim \mathcal{N} = r'$, and $\dim \mathcal{G} = d = r + r' + 2N$.

A typical element of $\mathcal{G}$ is of the form

$$X = \alpha_\mu \xi_\mu = v_\lambda R_\lambda + h_\mu \rho_\mu + \kappa_{\lambda\mu} \rho_{\lambda\mu} + \beta_\nu L_\nu$$

with $R_i \in \mathcal{P}$, $\rho_i \in \mathcal{H}$, $\rho_{ij} \in \mathcal{N}$, $L_i \in \mathcal{L}$. The corresponding factorization of the group is $\mathbf{G} = \mathbf{PKL} = \mathbf{PHNL}$, with group elements

$$e^X = e^{V_\lambda R_\lambda} e^{H_\mu \rho_\mu} \dots e^{C_{ij} \rho_{ij}} \dots e^{B_\lambda L_\lambda}$$

We denote group coordinates of $\mathbf{K}$ by A, when the refinement into $\mathbf{K} = \mathbf{HN}$ is not necessary to be made. For conciseness, we thus write

$$g(V, A, B) = e^{VR} e^{A\rho} e^{BL}$$

where, e.g., $VR = V_\mu R_\mu$, or when using double-indices, corresponding to bases for matrices, $VR = V_{\lambda\mu} R_{\lambda\mu}$, and similarly for $A\rho$ and BL.

1.1 ADJOINT ACTION

An element $\rho \in \mathcal{K}$ acts on $\mathcal{P}$ and $\mathcal{L}$ by $\operatorname{ad} \rho$. We have the adjoint matrix $\check{\rho}$ on the coordinates V:

$$(\operatorname{ad} \rho) V_\mu R_\mu = V_\mu \check{\rho}_{\lambda\mu} R_\lambda = (\check{\rho}_{\lambda\mu} V_\mu) R_\lambda$$
$$= (\check{\rho} V) R$$

where $\check{\rho}$, the matrix of ρ in the adjoint representation, is here restricted to $\mathcal{P}$. The action of $\mathcal{K}$ on $\mathcal{L}$ corresponds similarly to the adjoint action in terms of the coordinates B.

1.2 LEIBNIZ RULE

The translation group acting on functions of x, has generator D with group elements e^{aD} and the action $e^{aD} f(x) = f(x + a)$, on polynomials f, say. Now consider the Heisenberg group with Lie algebra generated by the boson operators X, D, where X denotes the operator of multiplication by x. We thus have

$$e^{aD} e^{bX} f(x) = e^{b(x+a)} f(x + a)$$

If we expand this in powers of a and b, using the Taylor expansion for f, we find for the coefficient of $(a^k/k!)(b^l/l!)$:

$$D^k X^l f(x) = \sum_j \frac{k!\, l!}{j!\,(k-j)!\,(l-j)!}\, x^{l-j} f^{(k-j)}(x)$$

which is Leibniz' rule for the kth derivative of the product $x^l f(x)$. That is, the group law encodes the formula for differentiating the product of functions.

In the present context, given the Cartan decomposition, the formula, defining $\tilde{V}, \tilde{A}, \tilde{B}$:

$$e^{BL} e^{VR} = g(0, 0, B)g(V, 0, 0) = g(\tilde{V}(B, V), \tilde{A}(B, V), \tilde{B}(B, V))$$

is the generating function for commuting L^n past R^m and generalizes the classical Leibniz rule. Thus, we call it the Leibniz rule for the group.

Given the Leibniz rule, the group law can be written out.

1.2.1 Proposition.   *The group law has the form*

$$g(V, A, B)g(V', A', B') = g(V + e^{A\hat{\rho}}\,\tilde{V}(B, V'), A \odot \tilde{A}(B, V') \odot A', e^{-A'\hat{\rho}}\,\tilde{B}(B, V') + B')$$

where the $\odot$ indicates the group law in the subgroup **K**.

Proof: Write out $g(V, A, B)g(V', A', B')$ and use the Leibniz rule:

$$e^{VR} e^{A\rho} e^{BL} e^{V'R} e^{A'\rho} e^{B'L}$$

$$= e^{VR} e^{A\rho} e^{\tilde{V}(B,V')R} e^{\tilde{A}(B,V')\rho} e^{\tilde{B}(B,V')L} e^{A'\rho} e^{B'L}$$

Now $e^{A\rho}$ acts on R by the adjoint representation of the group, i.e., the exponential of the adjoint representation of the algebra, so that

$$e^{A\rho} e^{\tilde{V}R} = \exp\!\left(e^{A\hat{\rho}}\,\tilde{V}\,R\right)e^{A\rho}$$

Similarly, $e^{A'\rho}$ acts on $e^{\tilde{B}L}$ from the right by the inverse adjoint action of the group. ∎

In summary: knowing the group law of **K** and the adjoint action of $\mathcal{K}$ on $\mathcal{P}$ and $\mathcal{L}$, the group law for **G** is determined.

1.3 GENERAL FORM OF THE DUAL REPRESENTATIONS

For algebras $\mathcal{G}$ with a Lie flag, we found a general form for the pi-matrices, Ch. 2, Theorem 2.2.1.1. Here, using the adjoint action of $\mathcal{K}$ on $\mathcal{P}$ and $\mathcal{L}$ we can give a general form of the dual representations. We assume that for $\mathcal{K}$, everything is known. Denote the dual representations of $\mathcal{K}$ itself, including the double dual, by $\rho_{\mathcal{K}}^{\dagger}$, $\hat{\rho}_{\mathcal{K}}$, $\rho_{\mathcal{K}}^{\bullet}$.

Now for $\mathcal{G}$. To find the left dual, first consider the left multiplication by R on $g(V, A, B)$. We have immediately the dual action $R_j^{\dagger} = \partial/\partial V_j$. Denote this by $R^{\dagger} = \partial_V$.

Now take $\rho \in \mathcal{K}$. We have

$$\rho e^{VR} = e^{VR}(e^{-VR}\rho e^{VR}) = e^{VR}(\rho + V(\operatorname{ad}\rho\, R))$$

where there are no further terms since $\operatorname{ad}\rho\, R \in \mathcal{P}$ and so commutes with R's. As is common in group theory, we concisely denote (in this section only) the adjoint action, here in the algebra, by $\operatorname{ad}\rho\, R = R^{\rho}$. Then ρ continues to the right, acting on $\mathbf{K}$, which we assume given: $\rho_{\mathcal{K}}^{\dagger}$. The R^{ρ} term dualizes to ∂_V^{ρ}. Thus,

$$\rho^{\dagger} = V\partial_V^{\rho} + \rho_{\mathcal{K}}^{\dagger}$$

For $L \in \mathcal{L}$,

$$Le^{VR} = e^{VR}(e^{-VR}Le^{VR}) = e^{VR}(L + V([L,R]) + \tfrac{1}{2}V^2([[L,R],R]))$$

where $[L,R] \in \mathcal{K}$ is denoted ρ_{LR}. Then

$$[[L,R],R] = [\rho_{LR},R] = R^{\rho_{LR}} \in \mathcal{R}$$

so that there are no further terms. The expression V^2 is symbolic for terms of the form V_iV_j that arise in the third term. That third term dualizes by replacing R by ∂_V as above. The second term gives V times the action of ρ_{LR} on $\mathcal{K}$, which is $\rho_{LR,\mathcal{K}}^{\dagger}$. Now, we must move L to the right past $e^{A\rho}$. This gives, via the adjoint action of $\mathcal{K}$,

$$Le^{A\rho}e^{BL} = e^{A\rho}e^{-A\operatorname{ad}\rho}Le^{BL}$$

Finally, we dualize L on e^{BL} to ∂_B. To get the double dual, we replace derivatives by the raising operators $\mathcal{R}$ and the variables by the velocity operators $\mathcal{V}$. The right dual is obtained analogously to the left dual.

Now we summarize the general forms:

For the left dual:

$$R^{\ddagger} = \partial_V$$
$$\rho^{\ddagger} = V\partial_V^{\rho} + \rho_{\kappa}^{\ddagger}$$
$$L^{\ddagger} = e^{-\text{Aad}\,\rho}\,\partial_B + V\rho_{LR,\kappa}^{\ddagger} + \tfrac{1}{2}V^2 R^{\rho LR}$$

The double dual has the form:

$$\hat{R} = \mathcal{R}$$
$$\hat{\rho} = \mathcal{R}^{\rho}\mathcal{V} + \hat{\rho}_{\kappa} \qquad\qquad (1.3.1)$$
$$\hat{L} = e^{-\text{Aad}\,\rho}\,\mathcal{B} + \hat{\rho}_{LR,\kappa}\mathcal{V} + \tfrac{1}{2}\mathcal{R}^{\rho LR}\mathcal{V}^2$$

For the right action:

$$R^* = e^{\text{Aad}\,\rho}\,\partial_V + B\rho_{LR,\kappa}^* - \tfrac{1}{2}B^2\partial_B^{\rho LR}$$
$$\rho^* = -B\partial_B^{\rho} + \rho_{\kappa}^*$$
$$L^* = \partial_V$$

These thus provide a sketch of the form of the pi-matrices in general.

1.4 INDUCED REPRESENTATION AND HOMOGENEOUS SPACE

Now consider the representation induced from the action on the vacuum Ω:

$$L_j\Omega = 0\,, \quad \rho_i\Omega = \tau_i\Omega\,, \quad \rho_{ij}\Omega = 0$$

with τ_i the weights of the representation. The action of the group is thus

$$g(V,A,B)\Omega = e^X\,\Omega = e^{\tau_\mu H_\mu}\,e^{V_\lambda R_\lambda}\,\Omega$$

The resulting homogeneous space has V for coordinates and is realized in the form $e^{VR}\,\Omega$. The group acts on the space via the group law. On $e^{V'R}\,\Omega$, we have the group action from Proposition 1.2.1:

$$g(V,A,B)g(V',0,0)\Omega = g(V + e^{A\hat{\rho}}\,\tilde{V}(B,V'), A \odot \tilde{A}(B,V'),0)\Omega \qquad (1.4.1)$$

cf., Ch. 2, Proposition 4.2.1.3. Thus

1.4.1 Proposition.   With $g(V,A,B)g(V',0,0)\Omega = e^{\tau H''}\,e^{V''R}\,\Omega$,

$$V'' = V + e^{A\hat{\rho}}\,\tilde{V}(B,V')\,, \qquad H'' = H + \tilde{H}(B,V')$$

where $A = (H,C), \tilde{A} = (\tilde{H},\tilde{C})$ according to the decomposition $\mathcal{K} = \mathcal{H} \oplus \mathcal{N}$.

Proof: The coefficients V'' are clear, as they are not affected by acting on Ω. From eq. (1.4.1),

$$g(0, A \odot \tilde{A}(B, V'), 0)\Omega = e^{A\rho} e^{\tau \tilde{H}(B,V')} \Omega = e^{\tau H} e^{\tau \tilde{H}(B,V')} \Omega$$

and hence the result. ∎

The induced representation of the algebra can be found from the double dual, eq. (1.3.1). The basis for the boson calculus is $[\![n]\!] = R^n \Omega$. The raising operators $\mathcal{R}_i$ are given by multiplication by R_i. The velocity operators $\mathcal{V}$ act as formal differentiation with respect to the R's. They act on $e^{VR}\Omega$ as multiplication by V. In $\hat{\rho}$ and $\hat{L}$, all of the variables corresponding to the ρ_{ij}'s map to zero, while those corresponding to the ρ_i map to scalars τ_i. In $\hat{L}$, all of the $\mathcal{B}$'s map to zero.

1.5 HARISH-CHANDRA EMBEDDING

Another way to study the group and the homogeneous space is to consider the Lie algebra/group in the form of block matrices. This is one general example of a Cartan decomposition that is useful for seeing some of the implications of those assumptions.

The basic model is to write $X \in \mathcal{G}$ in the block-matrix form

$$X = \begin{pmatrix} \alpha & v \\ -\beta & \delta \end{pmatrix} = vR + \beta L + K$$

with $K \in \mathcal{K}$. The diagonal block with entries α, δ comprise $K = h_\mu \rho_\mu + \kappa_{\lambda\mu} \rho_{\lambda\mu}$. The group elements take the form

$$g(V, E, D, B) = \begin{pmatrix} I & V \\ 0 & I \end{pmatrix} \begin{pmatrix} E & 0 \\ 0 & D \end{pmatrix} \begin{pmatrix} I & 0 \\ -B & I \end{pmatrix} = \begin{pmatrix} E - VDB & VD \\ -DB & D \end{pmatrix} \quad (1.5.1)$$

The embedding of the homogeneous space is given by how V occurs in the group elements. The main property is that

1.5.1 Proposition.   *Given g, we can recover V by the ratio of entries in the block matrix $g \begin{pmatrix} 0 \\ I \end{pmatrix}$.*

Proof: With g in the form given in eq. (1.5.1), we see that

$$g \begin{pmatrix} 0 \\ I \end{pmatrix} = \begin{pmatrix} VD \\ D \end{pmatrix}$$

and then V is found from the relation $V = (VD)D^{-1}$. ∎

In this sense, the V's are projective coordinates, and the representative form of $e^{VR}\Omega$ is $\begin{pmatrix} V \\ I \end{pmatrix}$.

1.6   GROUP COORDINATES

The group coordinates (V, E, D, B) are given in terms of $(v, \alpha, \delta, \beta)$ as in the splitting lemma, Ch. 2, 2.2.

1.6.1 Proposition.   Write $g(V(s), E(s), D(s), B(s)) = e^{sX}$, with $X = \begin{pmatrix} \alpha & v \\ -\beta & \delta \end{pmatrix}$.

Then

$$\dot{V} = v + \alpha V - V\delta + V\beta V$$
$$\dot{D} = (-\beta V + \delta)D$$
$$\dot{E} = (\alpha + V\beta)E$$
$$\dot{B} = D^{-1}\beta E$$

Proof: We have $\dot{g} = Xg$. First,

$$\dot{g} = \begin{pmatrix} \dot{E} - (VDB) & \dot{V}D + V\dot{D} \\ -\dot{D}B - D\dot{B} & \dot{D} \end{pmatrix}$$

while

$$Xg = \begin{pmatrix} \alpha E - \alpha VDB - vDB & \alpha VD + vD \\ -\beta E + \beta VDB - \delta DB & -\beta VD + \delta D \end{pmatrix}$$

Comparing yields the result.   ∎

Given a specific algebra, from these equations one can read off $\pi^{\dagger}$, according to the splitting lemma. Similar equations hold multiplying by X on the right which yield π^{*}.

Note that the flow for V is given by a matrix Riccati equation.

1.7   ORBITS

The basis for the induced representation is $R^n\Omega$. To see how the Lie algebra acts on this space, we consider the action of $X \in \mathcal{G}$ on $e^{VR}\Omega$, which is thought of as a generating function for the basis. We have

$$Xe^{VR}\Omega = e^{VR}(e^{-VR}Xe^{VR})\Omega$$

i.e., we can compute the adjoint orbit of $\mathbf{P}$ on X to find the action of X on the space. Using the matrices given above, we can compute directly.

1.7.1 Proposition.   For the adjoint action of $\mathbf{P}$ on X, we have the matrix form

$$e^{-VR}Xe^{VR} = \begin{pmatrix} \alpha + V\beta & v + \alpha V - V\delta + V\beta V \\ -\beta & -\beta V + \delta \end{pmatrix}$$

An interesting general result is the following

1.7.2 Theorem. Let $e^{sX}\,\Omega = e^{\tau H(s)}\,e^{V(s)R}\,\Omega$. Then

$$e^{-V(s)R}\,X e^{V(s)R}\,\Omega = (\tau\dot{H} + \dot{V}R)\Omega$$

the dot denoting differentiation with respect to s.

Proof: Differentiating the given relation,

$$X e^{sX}\,\Omega = e^{\tau H(s)}\,(\tau\dot{H} + \dot{V}R)e^{V(s)R}\,\Omega$$

Now, on the left side, rewrite $e^{sX}\,\Omega = e^{\tau H(s)}\,e^{V(s)R}\,\Omega$, cancel the common factor of $e^{\tau H(s)}$, and then multiply both sides by $e^{-V(s)R}$. Thus the result. ∎

In the light of this, compare the above Proposition with Proposition 1.6.1.

1.7.1 Group-theoretical solution of matrix Riccati equation

In our matrix formulation, we illustrate here the idea remarked in Ch. 2, §2.5, to solve for V using the group action. Observe that a generic X can be gotten from $\tilde{X} \in \mathcal{P} \oplus \mathcal{K}$ by the adjoint action of **L**:

$$e^{BL}\,\tilde{X}e^{-BL} = e^{BL}\begin{pmatrix} \alpha & v \\ 0 & \delta \end{pmatrix}e^{-BL}$$

$$= \begin{pmatrix} \alpha + vB & v \\ -B\alpha + \delta B - BvB & -Bv + \delta \end{pmatrix}$$

Now, the algebra $\mathcal{P} \oplus \mathcal{K}$ is of affine type, so we can use Ch. 2, Proposition 2.1.2, to conclude that $e^{s\tilde{X}} = e^{V_0(s)R}\,e^{sK}$ with

$$V_0(s)R = \int_0^s e^{u\operatorname{ad}_{\mathcal{P}}K}\,du\,vR \tag{1.7.1.1}$$

If $X = e^{BL}\,\tilde{X}e^{-BL}$, then

$$e^{sX} = e^{BL}\,e^{s\tilde{X}}\,e^{-BL} = \left(e^{BL}\,e^{V_0(s)R}\,e^{-BL}\right)\left(e^{BL}\,e^{sK}\,e^{-BL}\right) \tag{1.7.1.2}$$

with the second term of the product not involving R. In matrix terms we calculate

$$e^{BL}\,e^{UR}\,e^{-BL} = \begin{pmatrix} I & 0 \\ -B & I \end{pmatrix}\begin{pmatrix} I & U \\ 0 & I \end{pmatrix}\begin{pmatrix} I & 0 \\ B & I \end{pmatrix} = \begin{pmatrix} I+UB & U \\ -BUB & I-BU \end{pmatrix}$$

Thus, $U \to U(I - BU)^{-1}$ under the adjoint action of e^{BL}. And the solution to the Riccati equation has the form $V_0(I - BV_0)^{-1}$ where V_0 is given from eq. (1.7.1.1). To see what the situation is without using the block matrices, since we are only interested in the V coordinates, we can use the induced representation. I.e., apply both sides of eq. (1.7.1.2) to Ω. Ignoring terms from **KL**,

$$e^{BL}\,e^{V_0(s)R}\,\Omega = e^{\tau\tilde{H}(B,V_0)}\,e^{\tilde{V}(B,V_0)R}\,\Omega$$

which is given by the Leibniz rule.

1.8 SUMMARY OF FORMS

For reference, here is a summary of useful forms.

Group element:
$$e^{VR}\, e^{\rho H}\, e^{BL} = \begin{pmatrix} E - VDB & VD \\ -DB & D \end{pmatrix}$$

Inverse:
$$e^{-BL}\, e^{-\rho H}\, e^{-VR} = \begin{pmatrix} E^{-1} & -E^{-1}V \\ BE^{-1} & -BE^{-1}V + D^{-1} \end{pmatrix}$$

Action on induced representation, orbit of **P**:
$$e^{-VR}\, X e^{VR} = \begin{pmatrix} \alpha + V\beta & v + \alpha V - V\delta + V\beta V \\ -\beta & -\beta V + \delta \end{pmatrix}$$

Orbit of **L** on $\mathcal{P} \oplus \mathcal{K}$:
$$e^{BL}\, \tilde{X} e^{-BL} = \begin{pmatrix} \alpha + vB & v \\ -B\alpha + \delta B - BvB & -Bv + \delta \end{pmatrix}$$

Leibniz Rule:
$$e^{BL}\, e^{VR} =$$
$$\begin{pmatrix} I & V(I - BV)^{-1} \\ 0 & I \end{pmatrix} \begin{pmatrix} I + V(I - BV)^{-1}B & 0 \\ 0 & I - BV \end{pmatrix} \begin{pmatrix} I & 0 \\ -(I - BV)^{-1}B & I \end{pmatrix}$$

Action of **K** on $\mathcal{P}$:
$$e^{\rho H}\, vR e^{-\rho H} = \begin{pmatrix} 0 & EvD^{-1} \\ 0 & 0 \end{pmatrix}$$

Group Multiplication:
$$\begin{pmatrix} E - VDB & VD \\ -DB & D \end{pmatrix} \begin{pmatrix} E' - V'D'B' & V'D' \\ -D'B' & D' \end{pmatrix}$$
$$= \begin{pmatrix} * & EV'D' - VDBV'D' + VDD' \\ -DBE' + DBV'D'B' - DD'B' & -DBV'D' + DD' \end{pmatrix}$$

with $* = EE' - EV'D'B' - VDBE' + VDBV'D'B' - VDD'B'$.

   which gives for the group law $(V''', E''', D''', B''') = (V, E, D, B) \odot (V', E', D', B')$
$$V''' = V + EV'(I - BV')^{-1}D^{-1}, D''' = D(I - BV')D', B''' = B' + D'^{-1}(I - BV')^{-1}BE'$$
and
$$E''' = EE' + EV'(I - BV')^{-1}BE' = E[I + V'(I - BV')^{-1}B]E'$$

II.   Induced representation and coherent states

The *coherent states,* in the sense of Perelomov[47], are given by the action of a group element on the vacuum state, $e^X \Omega$. Since $e^X \Omega = e^{\tau H} e^{VR} \Omega$, we define the coherent states as

$$\psi_V = e^{VR} \Omega$$

scaling out $e^{\tau H}$. The coherent state is analytic in a neighborhood of $0 \in \mathbf{C}$, and so is the generating function for the basis $[\![n]\!] = R^n \Omega$.

We now assume that we can build an inner product $\langle\,,\,\rangle$ such that the operators L_i and R_i are mutually adjoint. The first requirement is that the coherent states have finite norm, as least in some neighborhood of 0: $\langle \psi_V, \psi_V \rangle < \infty$. Then we require

$$\langle R_i \psi_B, \psi_V \rangle = \langle \psi_B, L_i \psi_V \rangle \tag{2.1}$$

for all $1 \le i \le N$.

Note. We will do everything in general for real coefficients. For complex coefficients, bars indicating conjugation are to be inserted in appropriate places.

2.1 Definition.   The *Leibniz function* is defined by

$$\Upsilon_{BV} = \langle \psi_B, \psi_V \rangle$$

the inner product of the coherent states.

Using this as a normalizing factor, we define the coherent state transform of an operator

2.2 Definition.   The *coherent state representation* or *Berezin symbol* of Q is given by:

$$\langle Q \rangle_{BV} = \frac{\langle \psi_B, Q\psi_V \rangle}{\langle \psi_B, \psi_V \rangle}$$

Using the vacuum state,

2.3 Definition.   The *expected value* of Q is given by

$$\langle Q \rangle = \langle \Omega, Q\Omega \rangle$$

where Ω is normalized to $\|\Omega\| = 1$.

Some useful features are

2.4 Proposition.   *We have*

$$\langle R \rangle = \langle L \rangle = 0, \quad \langle \rho_i \rangle = \tau_i, \quad \langle e^{VR} \rangle = 1$$

Proof: Since $L\Omega = 0$, the result for L is clear, while

$$\langle R \rangle = \langle \Omega, R\Omega \rangle = \langle L\Omega, \Omega \rangle = 0$$

Since $\rho_i \Omega = \tau_i \Omega$, the equation for ρ_i follows from the normalization of Ω. As for R, from $\langle \Omega, e^{VR} \Omega \rangle$ bring e^{VR} across as e^{VL} which leaves just the norm square of Ω. $\blacksquare$

The main property is this:

2.5 Theorem.   *The Leibniz rule and the Leibniz function are related by:*

$$\Upsilon_{BV} = e^{\tau \tilde{H}(B,V)}$$

Proof: Starting from the left-hand side:

$$\langle \psi_B, \psi_V \rangle = \langle e^{BR} \Omega, e^{VR} \Omega \rangle = \langle \Omega, e^{BL} e^{VR} \Omega \rangle$$
$$= \langle \Omega, e^{\tau \tilde{H}(B,V)} e^{\tilde{V}(B,V)R} \Omega \rangle$$
$$= e^{\tau \tilde{H}(B,V)}$$

as in Proposition 2.4. $\blacksquare$

Observe that Υ_{BV} is the generating function for the inner products of the basis vectors:

$$\Upsilon_{BV} = \sum_{n,m} \frac{B^n V^m}{n!\, m!} \langle R^n \Omega, R^m \Omega \rangle \tag{2.2}$$

In other words, given the expansion $\Upsilon_{BV} = \sum (B^n/n!)(V^m/m!)\, c_{nm}$, the coefficients c_{nm} must satisfy the positivity conditions: given any finite set F of multi-indices

$$\sum_{m,n \in F} c_{nm} \zeta_n \zeta_m > 0 \tag{2.3}$$

for any numbers $\{ \zeta_n \}_{n \in F}$. This is called the *complete positivity* condition.

Another important feature:

2.6 Proposition.   *The Leibniz function Υ_{BV} satisfies the cocycle property*

$$\Upsilon_{B,U+V} = \Upsilon_{UB} \Upsilon_{\tilde{V}(U,B)V}$$

Proof: Start from $\Upsilon_{B,U+V}$:

$$\langle e^{BR}\Omega, e^{(U+V)R}\Omega\rangle = \langle e^{UL}e^{BR}\Omega, e^{VR}\Omega\rangle$$

$$= e^{\tau\tilde{H}(U,B)}\langle e^{\tilde{V}(U,B)R}, \psi_V\rangle$$

$$= e^{\tau\tilde{H}(U,B)}\Upsilon_{\tilde{V}(U,B)V}$$

and the result follows by Theorem 2.5. ∎

We summarize the properties of Υ_{BV}.

2.7 Proposition. *The Leibniz function Υ_{BV} has three essential properties:*

1. *Analyticity*

2. *Complete positivity*

3. *Cocycle property*

Given such a function, we will see later how to recover the representation of the Lie algebra in the induced representation. Note, however, that the cocycle property presupposes knowledge of the action of the group on the homogeneous space, the domain with coordinates V. The expansion of the Leibniz function determines the Hilbert space structure as well.

2.1 DISTRIBUTION OF X

If $X \in \mathcal{G}$ is self-adjoint — as an operator on the Hilbert space determined as above, with basis $R^n\Omega$ — then the unitary group generated by X satisfies

$$\langle e^{isX}\rangle = \langle\Omega, e^{isX}\Omega\rangle = \phi(s)$$

where $\phi(s)$ is a positive-definite function. Hence, Bochner's Theorem provides a probability measure $p(dx)$ such that

$$\phi(s) = \int_{-\infty}^{\infty} e^{isx}\, p(dx)$$

Thus, we can interpret X as a random variable with distribution p.

If we have a family $\{X_1, \ldots, X_q\}$ of mutually commuting self-adjoint operators, then again Bochner's Theorem applies and we have for the joint distribution

$$\langle e^{i\sum z_j X_j}\rangle = \int_{\mathbf{R}^q} e^{izx}\, p(dx)$$

The probability measures arising in this way, we say are *fundamental probability distributions.*

Remark. This has been illustrated in Volume 1, where we have seen that the Gaussian, Poisson, exponential, binomial, negative binomial, and continuous binomial distributions are fundamental not only heuristically and empirically, but in this sense as well, coming from representations of forms of sl(2) and contractions, specifically the oscillator and Heisenberg algebras.

We use the group action to find coherent states given those for elements of simple form on the same adjoint orbits. Suppose we have Y with $e^{sY} = e^{V(s)R} e^{A(s)\rho} e^{B(s)L}$. We generate X by the adjoint action of group elements on Y:

$$X = e^{B'L} e^{-V'R} Y e^{V'R} e^{-B'L}$$

For the coherent state generated by X:

$$e^{sX}\Omega = e^{B'L} e^{-V'R} e^{sY} e^{V'R} \Omega$$
$$= e^{B'L} e^{-V'R} g(V, A, B)g(V', 0, 0)\Omega$$
$$= e^{B'L} e^{\tau H''} e^{(V''-V')R} \Omega$$
$$= e^{\tau H} \Upsilon_{BV'} \Upsilon_{B', V''-V'} e^{\bar{V}(B', V''-V')R} \Omega$$

as in Proposition 1.4.1 and Theorem 2.5. Note that the factor $e^{\tau H}$ comes originally from the factorization of e^{sY} unchanged. We thus have

$$\langle e^{sX} \rangle = e^{\tau H} \Upsilon_{BV'} \Upsilon_{B', V''-V'}$$

If X is self-adjoint, then this gives the Fourier-Laplace transform of the distribution of X.

Example. Starting from $Y_i = R_i$, let $X_i = e^{BL} R_i e^{-BL}$, $1 \le i \le N$. This gives N commuting elements (not necessarily self-adjoint with respect to the given inner product). Then

$$e^{z_\mu X_\mu}\Omega = e^{BL} e^{zR}\Omega = \Upsilon_{Bz} e^{\bar{V}(B,z)R}\Omega \qquad (2.1.1)$$

and $\langle e^{z_\mu X_\mu} \rangle = \Upsilon_{Bz}$.

2.2 CONSTRUCTION OF THE ALGEBRA FROM COHERENT STATES

With the Lie algebra given, L_i acts on ψ_V by the double dual in the induced representation, cf. the remarks at the end of §1.4. Namely, all variables are set to zero except for those corresponding to the ρ_i, which map to scalars τ_i, and those for R_i. Thus, acting on ψ_V, $\hat{L}_i$ is given in terms of boson operators $\mathcal{R}, \mathcal{V}$.

Given a Leibniz function, Υ_{BV}, satisfying the requisite properties of analyticity, complete positivity, and the cocycle property for a given group law, writing $\Upsilon_{BV} =$

$\langle e^{BR}\,\Omega, e^{VR}\,\Omega \rangle$, we can construct the Lie algebra. First, we have $R_i\psi_V$ given by differentiating with respect to V_i (writing Υ with BV understood):

$$\frac{\partial \Upsilon}{\partial V_i} = \langle e^{BR}\,\Omega, R_i e^{VR}\,\Omega \rangle$$

Since we have L_i adjoint to R_i, we determine L_i by $\partial\Upsilon/\partial B_i$, which brings down R_i on the left, yielding $\langle \psi_B, L_i\psi_V \rangle$. So L_i will be given a boson realization, where, acting on ψ_V, $R_i = R_i$ and V_i corresponds to multiplication by V_i. Once the boson realization of the L_i is known, the Lie algebra generated by L and R can be found. We have

2.2.1 Proposition.   *The Leibniz function Υ_{BV} satisfies the differential equations*

$$\frac{\partial \Upsilon}{\partial B_i} = \hat{L}_i \Upsilon$$

where $\hat{L}_i$ is the double dual of L_i in the induced representation, and R_k is given by $\frac{\partial}{\partial V_k}$.

We give a simple illustration.

Example. The function, for $t > 0$,

$$\Upsilon_{BV} = e^{tBV} = e^{tB_\mu V_\mu}$$

is a Leibniz function. We have

$$\frac{\partial \Upsilon}{\partial B_i} = tV_i \Upsilon$$

and thus $\hat{L}_i = tV_i$ and we have the commutation relations of the Heisenberg algebra: $[L_i, R_j] = t\delta_{ij}I$. We have seen the Leibniz rule formulated in terms of D and X in §1.2. Here it takes the form

$$e^{BL}\, e^{VR} = e^{VR}\, e^{tBV}\, e^{BL}$$

or, $\tilde{V}(B,V) = V$, $\tilde{H}(B,V) = BV$. The cocycle property $\Upsilon_{B,U+V} = \Upsilon_{UB}\Upsilon_{\tilde{V}(U,B)V}$ is readily verified.

2.3 CANONICAL APPELL SYSTEMS

Write $e^{sX}\,\Omega = e^{\tau H(s)}\, e^{V(s)R}\,\Omega$. Consider the one-dimensional case, i.e., $N = 1$, with $\tau = t$ so that

$$e^{V(s)R}\,\Omega = e^{-tH(s)}\, e^{sX}\,\Omega$$

which is the form for the Appell system with $V(D)$ as canonical velocity operator. Write Theorem 1.7.2 in the form

$$Xe^{V(s)R}\,\Omega = (t\dot{H} + \dot{V}R)e^{V(s)R}\,\Omega$$

As operators on $e^{V(s)R}\Omega$, $X = t\dot{H} + \dot{V}R$ or $R = (X - t\dot{H})\dot{V}^{-1}$ as in Ch. 1, except here with $t \to -t$.

For $N > 1$, we want $\{X_1, \ldots, X_N\}$ commuting operators. Then

$$e^{z_\mu X_\mu}\Omega = e^{\tau H(z)}e^{V(z)R}\Omega$$

with $V(z)$ locally analytic, $V(0) = 0$. Now take z_i as parameter. Since the X_i commute,

$$X_i e^{z_\mu X_\mu}\Omega = \frac{\partial}{\partial z_i}e^{z_\mu X_\mu}\Omega$$

$$= \left(\tau\frac{\partial H}{\partial z_i} + \frac{\partial V_\mu}{\partial z_i}R_\mu\right)e^{z_\mu X_\mu}\Omega$$

For nondegenerate V, i.e., the Jacobian $V'(0)$ invertible, we have exactly the canonical Appell system with $Y_i = X_\lambda W_{\lambda i}$, $W = (V')^{-1}$. And R_i are raising operators for the system. I.e., $R_i = \left(e^{-\tau H}Xe^{\tau H}\right)_\lambda W_{\lambda i}$, so that the H_i are Hamiltonians for the flow generated by $-\tau_i$.

2.3.1 Proposition. *For commuting operators $X_i = e^{BL}R_i e^{-BL}$, the canonical velocity operator has symbol $\tilde{V}(B, z)$ and the Hamiltonians generating the flow with parameters $-\tau_i$ have symbols $\tilde{H}_i(B, z)$.*

Proof: This follows from the example given in §2.1, eq. (2.1.1):

$$e^{z_\mu X_\mu}\Omega = e^{\tau\tilde{H}(B,z)}e^{\tilde{V}(B,z)R}\Omega$$

and the discussion above applies. ∎

For the self-adjoint case, we have $\langle e^{z_\mu X_\mu}\rangle = e^{\tau H(z)}$. This gives the Fourier-Laplace transform of the joint distribution of $(X_1, \ldots, X_N)$, in other words the joint spectral measure (in the vacuum state). Furthermore, we have the relation between the Hamiltonian(s) H and the Leibniz function. Here we write zX and zx for $z_\lambda X_\lambda$ and $z_\lambda x_\lambda$ respectively.

2.3.2 Proposition. *Let $e^{zX}\Omega = e^{\tau H(z)}e^{V(z)R}\Omega$. Let X_j, $1 \leq j \leq N$, be realized as multiplication by x_j, $1 \leq j \leq N$, $\Omega = 1$, so that we have the canonical Appell system with raising operators R and canonical velocity operator with symbol $V(z)$. Then the Leibniz function satisfies*

$$\log\Upsilon_{V(z)V(z')} = \tau\left(H(z + z') - H(z) - H(z')\right)$$

Proof: We have $\langle e^{zX}\rangle = e^{\tau H(z)}$. Thus,

$$\langle e^{zx-\tau H(z)}e^{z'x-\tau H(z')}\rangle = \exp\left[\tau\left(H(z+z') - H(z) - H(z')\right)\right]$$

on the one hand, and on the other,

$$= \langle e^{V(z)R}1, e^{V(z')R}1\rangle = \Upsilon_{V(z)V(z')}$$

∎

III. Orthogonal polynomials in several variables

Here we begin the theme of the remainder of the volume: to study properties of Lie algebras and associated Lie groups by the theory presented in these first four chapters.

3.1 SL(N+1)

The Lie algebra $sl(N+1)$ of $(N+1) \times (N+1)$ matrices of trace zero we write in the block form

$$X = \begin{pmatrix} \alpha & v \\ -\beta^\dagger & \delta \end{pmatrix}$$

where v, β are $N \times 1$, α is $N \times N$ and δ is 1×1. We have $\delta = -\operatorname{tr}\alpha$.

The decomposition of $\mathcal{G}$ is given as follows. The R_i correspond to the entries in column $N+1$. I.e., in terms of boson operators via the Jordan map, $R_i = \mathcal{R}_i \mathcal{V}_{N+1}$. L_i is the negative transpose of R_i, so that $L_i = -\mathcal{R}_{N+1} \mathcal{V}_i$. The rest of the basis belongs to $\mathcal{K}$.

In $\mathcal{K}$, the diagonal elements — the Cartan elements — form an N-dimensional subspace. Denote a basis by $\rho_0, \rho_1, \ldots, \rho_{N-1}$. For computational purposes it is useful to have an orthogonal basis with respect to the standard inner product on matrices: $\langle M_1, M_2 \rangle = \operatorname{tr}(M_1^\dagger M_2)$. We build the ρ_i, $1 \le i < N$ inductively in N. We describe α_i, the upper left corner of ρ_i, as $\delta = 0$, for all of the ρ_i, $i \ne 0$. For $N = 2$, $\alpha_1 = \begin{pmatrix} 1 & 0 \\ 0 & -1 \end{pmatrix}$. Inductively, $\alpha_{N-1} = \operatorname{diag}(1-N, 1, \ldots, 1)$, the diagonal matrix with the indicated entries. Then $\alpha_2, \alpha_3, \ldots$ are of the form $\operatorname{diag}(0, \alpha_2^{(N-1)})$, $\operatorname{diag}(0, \alpha_3^{(N-1)})$, $\ldots$, where $\alpha_2^{(N-1)}$, $\alpha_3^{(N-1)}, \ldots$, are from the next lower dimension. Explicitly,

$$N = 3: \qquad \alpha_2 = \begin{pmatrix} -2 & 0 & 0 \\ 0 & 1 & 0 \\ 0 & 0 & 1 \end{pmatrix}, \qquad \alpha_1 = \begin{pmatrix} 0 & 0 & 0 \\ 0 & 1 & 0 \\ 0 & 0 & -1 \end{pmatrix}$$

$$N = 4: \qquad \alpha_3 = \begin{pmatrix} -3 & 0 & 0 & 0 \\ 0 & 1 & 0 & 0 \\ 0 & 0 & 1 & 0 \\ 0 & 0 & 0 & 1 \end{pmatrix}, \quad \alpha_2 = \begin{pmatrix} 0 & 0 & 0 & 0 \\ 0 & -2 & 0 & 0 \\ 0 & 0 & 1 & 0 \\ 0 & 0 & 0 & 1 \end{pmatrix}, \quad \alpha_1 = \begin{pmatrix} 0 & 0 & 0 & 0 \\ 0 & 0 & 0 & 0 \\ 0 & 0 & 1 & 0 \\ 0 & 0 & 0 & -1 \end{pmatrix}$$

and so on.

Now we define the special element ρ_0 by $\alpha = I$, $\delta = -N$. Then

3.1.1 Proposition. *The element ρ_0 satisfies:*

1. $[\rho_0, K] = 0$, $\forall K \in \mathcal{K}$.

2. $[\rho_0, R_i] = (1 + N)R_i$, $\quad [L_i, \rho_0] = (1 + N)L_i$

Proof: ρ_0 commutes with $\mathcal{K}$ since α is the identity and δ is a scalar. Compute using block matrices:

$$\left[\begin{pmatrix} I & 0 \\ 0 & -N \end{pmatrix}, \begin{pmatrix} 0 & v \\ -\beta^\dagger & 0 \end{pmatrix}\right] = \begin{pmatrix} 0 & (1+N)v \\ (1+N)\beta^\dagger & 0 \end{pmatrix}$$

from which we read off the result. ∎

3.1.2 Corollary. *The Cartan decomposition $\mathcal{G} = \mathcal{P} \oplus \mathcal{K} \oplus \mathcal{L}$ is the eigenspace decomposition of $\mathrm{ad}\,\rho_0$, with eigenvalues $1 + N$, 0, and $-(1 + N)$ respectively.*

This is a general feature of Hermitian symmetric spaces that will appear in Chapter 7 as well.

The ρ_{ij} correspond to the off-diagonal entries of α. They can be generated by bracketing L's with R's. Using boson operators,

$$[L_j, R_i] = [\mathcal{R}_i \mathcal{V}_{N+1}, \mathcal{R}_{N+1} \mathcal{V}_j]$$
$$= \mathcal{R}_i \mathcal{V}_j - \delta_{ij} \mathcal{R}_{N+1} \mathcal{V}_{N+1}$$

Consequently,

3.1.3 Proposition. *We have the relation: $\sum_i [L_i, R_i] = \rho_0$.*

The induced representation is given according to

$$L_i \Omega = 0, \quad \rho_{ij} \Omega = 0, \quad \rho_i \Omega = 0, 1 \leq i < N, \quad \rho_0 \Omega = tN\Omega$$

so that

$$g\Omega = e^{VR} e^{\rho_0 H_0} \Omega = e^{NtH_0} e^{VR} \Omega \tag{3.1.1}$$

where H_0 is the coordinate corresponding to ρ_0. In block matrix form,

$$g = \begin{pmatrix} I & V \\ 0 & I \end{pmatrix} \begin{pmatrix} E & 0 \\ 0 & D \end{pmatrix} \begin{pmatrix} I & 0 \\ -B^\dagger & I \end{pmatrix} \tag{3.1.2}$$

From §1.8, the Leibniz rule reads:

$$e^{BL} e^{VR} =$$
$$\begin{pmatrix} I & V(I - B^\dagger V)^{-1} \\ 0 & I \end{pmatrix} \begin{pmatrix} I + V(I - B^\dagger V)^{-1} B^\dagger & 0 \\ 0 & I - B^\dagger V \end{pmatrix} \begin{pmatrix} I & 0 \\ -(I - B^\dagger V)^{-1} B^\dagger & I \end{pmatrix} \tag{3.1.3}$$

3.1.4 Lemma. *For g of the form $\begin{pmatrix} * & V D \\ * & D \end{pmatrix}$,*

$$e^{H_0} = D^{-1/N}$$

Proof: From eq. (3.1.2) we see that the only factor contributing to D is $e^{\rho_0 H_0}$. And

$$e^{\rho_0 H_0} = \begin{pmatrix} e^{H_0} & 0 \\ 0 & e^{-N H_0} \end{pmatrix}$$

so that $D = e^{-N H_0}$ and hence the result. ∎

Thus,

3.1.5 Theorem. *For* $sl(N+1)$*, a Leibniz function is given by*

$$\Upsilon_{BV} = (1 - B^\dagger V)^{-t}$$

with $\tilde{V}(B,V) = V(1 - B^\dagger V)^{-1}$

Proof: From general theory, the Leibniz function satisfies

$$e^{BL} e^{VR} \Omega = \Upsilon_{BV} e^{\tilde{V}(B,V)R} \Omega$$

with $\Upsilon_{BV} = e^{t\tilde{H}_0}$. By the Lemma,

$$e^{\rho_0 \tilde{H}_0} \Omega = e^{Nt\tilde{H}_0} \Omega = D^{-t}$$

while from eq. (3.1.3), $D = 1 - B^\dagger V$. Hence the result for Υ. $\tilde{V}$ can be read off from eq. (3.1.3) directly.

Now we check the cocycle identity, Proposition 2.6. Taking logarithms and dividing out the common factor $-t$:

$$\log(1 - B^\dagger(U + V)) = \log(1 - U^\dagger B) + \log(1 - (1 - B^\dagger U)^{-1} B^\dagger V)$$

which is clear.

Complete positivity is seen from the expansion

$$(1 - B^\dagger V)^{-t} = \sum_{k=0}^{\infty} \frac{(t)_k}{k!} (B_\mu V_\mu)^k \tag{3.1.4}$$

so that the inner product is diagonal in the basis $R^n \Omega$, with positive squared norms. ∎

3.1.6 Corollary. *The Hilbert space has orthogonal basis* $\psi_n = R^n \Omega$ *and squared norms*

$$\gamma_n = \langle \psi_n, \psi_n \rangle = n! \, (t)_{|n|}$$

Proof: In the expansion eq. (3.1.4), by multinomial expansion

$$(B_\mu V_\mu)^k = k! \sum_{|n|=k} \prod \frac{(B_i V_i)^{n_i}}{n_i!}$$

With γ_n the coefficient of $\prod (B_i^{n_i}/n_i!)(V_i^{n_i}/n_i!)$, filling in the additional factor of $n!$, the result follows. ∎

To reconstruct the algebra from the Leibniz function $\Upsilon = \Upsilon_{BV}$, we differentiate to find:

$$\frac{1}{\Upsilon} \frac{\partial \Upsilon}{\partial B_i} = t \frac{V_i}{1 - B_\mu V_\mu}$$

$$\frac{1}{\Upsilon} \frac{\partial \Upsilon}{\partial V_i} = t \frac{B_i}{1 - B_\mu V_\mu} \tag{3.1.5}$$

so that

$$\frac{\partial \Upsilon}{\partial B_i} = t V_i \Upsilon + V_i \frac{\partial \Upsilon}{\partial V_\mu} V_\mu$$

This gives immediately, identifying R_i with $\partial \Upsilon / \partial V_i$ and multiplication by V_i with $\mathcal{V}_i$,

3.1.7 Theorem. *In terms of boson variables $\{ \mathcal{R}_i, \mathcal{V}_i \}_{1 \le i \le N}$ we have for the Lie algebra in the induced representation*

$$R_i = \mathcal{R}_i, \qquad L_i = t \mathcal{V}_i + \mathcal{R}_\mu \mathcal{V}_\mu \mathcal{V}_i$$

Note that L_i is of the form $(t + \text{number op.})\mathcal{V}_i$. And

$$[L_j, R_i] = \delta_{ij}(t + \mathcal{R}_\mu \mathcal{V}_\mu) + \mathcal{R}_i \mathcal{V}_j \tag{3.1.6}$$

And,

3.1.8 Corollary. ρ_0 *is given by* $Nt + (N+1)\mathcal{R}_\mu \mathcal{V}_\mu$.

Proof: Proposition 3.1.3 gives the relation $\rho_0 = \sum [L_i, R_i]$. Combining this with eq. (3.1.6) yields the result. ∎

I.e., ρ_0 is a scalar (multiple of the identity) plus a multiple of the number operator. This form of ρ_0 show its properties quite clearly: $\rho_0 \Omega = Nt\Omega$, and the commutation relations $[\rho_0, R_i] = (N+1)R_i$, $[L_i, \rho_0] = (N+1)L_i$, $[\rho_0, [L_j, R_i]] = 0$, $\forall 1 \le i, j \le N$.

Another important consequence of Lemma 3.1.4 connects H and V.

3.1.9 Proposition. Let $X = \begin{pmatrix} \alpha & v \\ -\beta\dagger & \delta \end{pmatrix}$, $e^{sX}\Omega = e^{tH(s)} e^{V(s)R} \Omega$. Then

$$\dot{H} = \beta^\dagger V(s) - \delta$$

Proof: From eq. (3.1.1), $H(s) = NH_0(s)$. By Lemma 3.1.4 and Proposition 1.6.1,

$$\dot{H}_0 = -\frac{1}{N}\frac{\dot{D}}{D} = \frac{1}{N}(\beta^\dagger V(s) - \delta)$$

Multiplying both sides by N yields the result. ∎

3.2 MULTINOMIAL DISTRIBUTIONS

The multinomial distribution arises in repeated independent trials. Suppose there are N possible outcomes per trial, with probabilities p_i, $1 \leq i \leq N$, with the additional possibility that none of these N outcomes occur, having probability $p_0 = 1 - \sum p_i$. Let X_i, $1 \leq i \leq N$, be the number of times outcome i has occurred in t trials; here $t > 0$ is restricted to the integers. Then

$$P(X_1 = n_1, \ldots, X_N = n_N) = \frac{N!}{n_0! \, n!} p_0^{t-|n|} p_1^{n_1} \cdots p_N^{n_N}$$

where indices run from 1 to N and the multi-indices $n = (n_1, \ldots, n_N)$. We make the nondegeneracy assumption that $p_0 > 0$ and all $p_i > 0$, $1 \leq i \leq N$.

The Fourier-Laplace transform is

$$\langle e^{z_\mu X_\mu} \rangle = \left(\sum_{j=1}^N p_j(e^{z_j} - 1) + 1\right)^t = e^{tH(z)}$$

Thus, the canonical Appell system has the form

$$e^{z_\mu X_\mu - tH(z)} = \sum_{n \geq 0} \frac{V(z)^n}{n!} [\![n]\!] \tag{3.2.1}$$

Now think of the process as a random walk on the lattice $\mathbf{Z}^N$ such that at each step the process jumps in direction $\mathbf{e}_i$ with probability p_i or sits, with probability p_0. Then we can introduce an orientation matrix C corresponding to the direction of possibile jumps, rescale and center, so that $H'(0) = 0$:

$$\langle e^{z_\mu X_\mu} \rangle = \left(\sum_k p_k(\exp(p_k^{-1}C_{k\lambda}z_\lambda) - 1) + 1\right)^t e^{-t\sum_k C_{k\lambda}z_\lambda} \tag{3.2.2}$$

An appropriate choice of C will lead to a Leibniz function and basis of orthogonal polynomials. First we need a technical observation.

3.2.1 Proposition. *Let $p = \text{diag}(p_1, \ldots, p_N)$ and denote the matrix having entries all ones by $\mathcal{O}$. Then $p^{-1} - \mathcal{O}$ is positive definite symmetric.*

Proof: We must check that in the standard Euclidean inner product, for non-zero $\mathbf{v} = (v_1, \ldots, v_N)$,

$$\langle \mathbf{v}, p^{-1}\mathbf{v} \rangle > \langle \mathbf{v}, \mathcal{O}\mathbf{v} \rangle = \left(\sum v_i \right)^2$$

Writing on the right-hand side, $v_i = (v_i/\sqrt{p_i})\sqrt{p_i}$, the result follows by Cauchy-Schwarz' inequality and the fact that $\sum p_i = 1 - p_0 < 1$. $\blacksquare$

From this Proposition it follows that we can define Q by

$$p^{-1} - \mathcal{O} = Q^\dagger Q \qquad (3.2.3)$$

And we set $C = Q^{-1}$. For convenience:

Notation. The vector $u = (u_1, \ldots, u_N)^\dagger$ has components all equal to 1.

Thus, as in eq. (3.2.2) we have $H(z)$ in the form

$$H(z) = \log \left(\sum_k p_k \left(\exp(p_k^{-1} C_{k\lambda} z_\lambda) - 1 \right) + 1 \right) - u^\dagger C z \qquad (3.2.4)$$

We need some useful identities that follow from eq. (3.2.3).

3.2.2 Proposition.   For any $a = (a_1, \ldots, a_N)^\dagger$, $b = (b_1, \ldots, b_N)^\dagger$:

1. $a_\lambda Q_{\lambda\mu} p_\mu = p_0 \, u^\dagger C a$

2. $a_\lambda Q_{\lambda\mu} b_\varepsilon Q_{\varepsilon\mu} p_\mu = a^\dagger b - p_0 (u^\dagger C a)(u^\dagger C b)$.

Proof: Here we indicate the steps of the calculations. For the first identity, multiply eq. (3.2.3) by $C^\dagger$ to get

$$Q = C^\dagger(p^{-1} - \mathcal{O}) \qquad (3.2.5)$$

The identity follows from this equation by contracting with a on the left and p on the right. For the second identity, multiply eq. (3.2.5) on the right by C to get:

$$I = C^\dagger(p^{-1} - \mathcal{O})C \qquad (3.2.6)$$

Now contract eq. (3.2.5) on the left, first by a and then by b, separately. Multiply these two equations and then contract with p on the right. Using eq. (3.2.6) the result simplifies as indicated. $\blacksquare$

With $H(z)$ given by eq. (3.2.4), define

$$\Delta = \sum_k p_k \left(\exp(p_k^{-1} C_{k\lambda} z_\lambda) - 1 \right) + 1$$

so that $H = \log \Delta$ − centering terms. As suggested by Proposition 3.1.9, we set, for the canonical Appell structure, eq. (3.2.1),

$$V_i(z) = \frac{\partial H}{\partial z_i}$$

or in brief, $V(z) = H'(z)$. For calculations with V it is convenient to set

$$E_k = \exp(p_k^{-1} C_{k\lambda} z_\lambda), \qquad \Delta = p_0 + p_\mu E_\mu \qquad (3.2.7)$$

Then

3.2.3 Proposition.   With $V_i(z) = \partial H/\partial z_i$, $1 \le i \le N$,
1. $\Delta^{-1} = 1 - p_0^{-1} V_\lambda Q_{\lambda\mu} p_\mu$
2. $\Delta^{-1} = 1 - u^\dagger C V$.

Proof: Directly from eq. (3.2.4), differentiation yields:

$$V_i = \frac{E_\mu C_{\mu i}}{\Delta} - u_\lambda C_{\lambda i}$$

Multiplying by Q, we have the relation

$$V_\lambda Q_{\lambda i} = \frac{E_i}{\Delta} - 1 \tag{3.2.8}$$

Contracting with p, i.e., multiplying by p_i and summing, yields the first relation after some rearrangement, via eq. (3.2.7) and the fact that $p_0 = 1 - \sum p_i$. The second relation follows from the first using #1 of Proposition 3.2.2.   ∎

For the canonical Appell systems, we have the flows corresponding to differentiation by each of the z_k. In particular, the V's satisfy Riccati equations. Thus,

3.2.4 Proposition.   *The Riccati equation holds in the form*

$$\frac{\partial V_j}{\partial z_k} = \delta_{jk} + a_{jk}^\lambda V_\lambda - V_j V_k$$

with the a-coefficients given by either of the expressions
1. $a_{ik}^j = p_\mu^{-1} C_{\mu i} Q_{j\mu} C_{\mu k} - u_\lambda C_{\lambda i} \delta_{kj} - u_\lambda C_{\lambda k} \delta_{ij}$
2. $a_{ik}^j = Q_{i\mu} Q_{j\mu} C_{\mu k} - u_\lambda C_{\lambda k} \delta_{ij}$.

These follow by direct calculation, eq. (3.2.3) mediates between the two forms.

We want to obtain the form of the canonical Appell system

$$e^{zU(v) - tH(U(v))} = \sum_{n \ge 0} \frac{v^n}{n!} [\![n]\!] \tag{3.2.9}$$

where U denotes the inverse function to V: $z = U(V)$. From eq. (3.2.7),

$$U_j(v) = \log \prod_k E_k^{Q_{jk} p_k}$$

Combining this with eq. (3.2.8), rewriting E in terms of Δ,

3.2.5 Proposition.

$$U_j(v) = \log\left[\prod_k \Delta^{Q_{jk}p_k}(1 + v_\lambda Q_{\lambda k})^{Q_{jk}p_k}\right]$$

And

3.2.6 Theorem.   *The generating function for the Appell system is*

$$e^{xU(v)-tH(U(v))} = (1 - u^\dagger Cv)^{p_0 t - x_\lambda Q_{\lambda\mu}p_\mu} \cdot \prod_k (1 + v_\epsilon Q_{\epsilon k})^{p_k(x_\alpha Q_{\alpha k}+t)}$$

Proof: We use eq. (3.2.9). From Proposition 3.2.5 we find $e^{xU(v)}$. For the factor $e^{-tH(U(v))}$, cf. eq. (3.2.4),

$$e^{-tH(U(v))} = \Delta^{-t} e^{tu_\lambda C_{\lambda\mu}U(v)_\mu}$$

Again use Proposition 3.2.5 and substitute for Δ by #2 of Proposition 3.2.3 to find the result stated. ∎

Now to find more about the form of the basis, we observe a connection with multivariate hypergeometric functions. Recall the *Lauricella polynomials*

$$F_B\left(\begin{matrix}-m, b\\ t\end{matrix} \,\bigg|\, \zeta\right) = \sum \frac{(-m)_n(b)_n \zeta^n}{(t)_{|n|}n!}$$

with N-component m, b, and ζ, t a number, $(-m)_n = \prod_{j=1}^N (-m_j)_{n_j}$, etc. These have the generating function

$$\left(1 - \sum v_i\right)^{u_\lambda b_\lambda - t} \prod_j \left(1 - \sum v_i + \zeta_j v_j\right)^{-b_j} = \sum \frac{v^m(t)_{|m|}}{m!} F_B\left(\begin{matrix}-m, b\\ t\end{matrix} \,\bigg|\, \zeta\right) \quad (3.2.10)$$

We thus have

3.2.7 Proposition.   *The basis $[\![n]\!]$ satisfies*

$$\sum \frac{v^n[\![n]\!]}{n!} = \sum \frac{(Cv)^m}{m!}(-t)_{|m|} F_B\left(\begin{matrix}-m, -pt - y\\ -t\end{matrix} \,\bigg|\, p^{-1}\right)$$

where $y_i = p_i x_\lambda Q_{\lambda i}$.

Proof: Substitute $v \to Cv$, $b_i \to -p_i(t + x_\lambda Q_{\lambda i})$, $\zeta_i \to p_i^{-1}$, $t \to -t$ in eq. (3.2.10) and compare with Theorem 3.2.6. ∎

Next we will calculate the Leibniz function, using Proposition 2.3.2:

$$H(U(a) + U(b)) - H(U(a)) - H(U(b)) = t^{-1} \log \Upsilon_{ab}$$

Note that centering terms are linear and drop out. Thus,

$$H(U(a) + U(b)) - H(U(a)) - H(U(b)) = \log \frac{\Delta(U(a) + U(b))}{\Delta(U(a))\Delta(U(b))}$$

3.2.8 Proposition.   *The Leibniz function is given by*

$$\Upsilon_{BV} = (1 + B^\dagger V)^t$$

Proof: Recalling eq. (3.2.8), $E_k = \Delta(1 + V_\lambda Q_{\lambda k})$. Thus, $E_k(U(a)) = \Delta(a)(1 + a_\lambda Q_{\lambda k})$. And since E is an exponential, $E(U(a) + U(b)) = E(U(a)) \cdot E(U(b))$. Therefore, via $\Delta = p_0 + p_\mu E_\mu$, eq. (3.2.7),

$$\frac{\Delta(U(a) + U(b))}{\Delta(U(a))\Delta(U(b))} = \frac{p_0 + p_\mu \Delta(U(a))\Delta(U(b))(1 + a_\lambda Q_{\lambda\mu})(1 + b_\varepsilon Q_{\varepsilon\mu})}{\Delta(U(a))\Delta(U(b))}$$

$$= \frac{p_0}{\Delta(U(a))\Delta(U(b))} + p_\mu(1 + a_\lambda Q_{\lambda\mu})(1 + b_\varepsilon Q_{\varepsilon\mu})$$

Now rewrite the fraction using Proposition 3.2.3:

$$\frac{p_0}{\Delta(U(a))\Delta(U(b))} = p_0(1 - u^\dagger Ca)(1 - u^\dagger Cb)$$

Substitute this into the above equation. Multiplying out, and using Proposition 3.2.2 to convert the Q expressions in terms of C, only $a^\dagger b$ remains and hence the result. ∎

As in Corollary 3.1.6, expanding the Leibniz function yields

3.2.9 Corollary.   *The Hilbert space has orthogonal basis $\psi_n = R^n \Omega$ and squared norms*

$$\gamma_n = n! \binom{t}{|n|} |n|!$$

Thus, since t is an integer, for each t, the representation is finite-dimensional.

The finite-dimensionality comes about by observing that basis vectors with $|n| \geq t$ have norm zero.

Now we can find the boson realization of the Lie algebra as in Proposition 2.2.1, cf. Theorem 3.1.7. As in eqs. (3.1.5), which show Theorem 3.1.7, we find

3.2.10 Proposition. *The Leibniz function Υ satisfies*

$$\frac{\partial \Upsilon}{\partial B_i} = t V_i \Upsilon - V_i \frac{\partial \Upsilon}{\partial V_\mu} V_\mu$$

so that the lowering operators have the form

$$L_i = t V_i - \mathcal{R}_\mu V_\mu V_i$$

And the Lie algebra is generated by the R's and L's.

3.2.11 Corollary. *The Lie algebra is a form of sl$(N+1)$. In particular,*
$\rho_0 = Nt - (N+1)\mathcal{R}_\mu V_\mu$.

Notice that this differs from the result in §3.1, Theorem 3.1.7, by a change of sign, as seen above in the form of the Leibniz function. This form of the algebra corresponds to the compact real form $su(N+1)$.

3.3 ORTHOGONALITY AND CONVOLUTION

We conclude with a sketch of a 'group theory' construction, using Fourier-Laplace transform and convolution that builds an orthogonal system from a given one. It is closely related to the reduction of the tensor product of two copies of the given L^2 space as in the construction of Clebsch-Gordan coefficients (cf., Chapter 3 of Volume 1).

Remark. In this section, unless otherwise indicated, we will discuss the $N = 1$-dimensional case, for convenience. The constructions indicated hold for $N > 1$ as well, appropriately modified.

Orthogonal polynomials (in one variable) may be described in terms of Fourier-Laplace transforms as follows. Given a measure $p(dx)$, the functions $\phi_n(x)$ are orthogonal to all polynomials of degree less than n if and only if

$$\int_{-\infty}^{\infty} e^{sx}\, \phi_n(x)\, p(dx) = V_n(s)$$

such that $V_n(s)$ has a zero of order n at $s = 0$. The proof is immediate from

$$\int_{-\infty}^{\infty} x^k \phi_n(x)\, p(dx) = \left(\frac{d}{ds}\right)^k \bigg|_0 V_n(s)$$

Thus, if the $\phi_n(x)$ are polynomials, they form a sequence of orthogonal polynomials.

3.3.1 Convolutions and Orthogonal functions

Start with a family of functions, *kernels,*

$$K(x, z, A)$$

where A indicates some parameters, that form a group under convolution

$$\int_{-\infty}^{\infty} K(x - y, z, A)K(y, z', A')\, dy = K(x, z + z', A'')$$

(The integration here can be replaced analogously by a summation.) This means that the Fourier-Laplace transforms form a multiplicative family. Let

$$\hat{K}(s, z, A) = \int_{-\infty}^{\infty} e^{sy}\, K(y, z, A)\, dy$$

Then

$$\hat{K}(s, z, A) \times \hat{K}(s, z,' A') = \hat{K}(s, z + z', A'')$$

Form the product

$$K(x - y, -z, A)K(y, z, A')$$

This integrates to $K(x, 0, A'')$ which is independent of z. This is the generating function for the orthogonal functions we are looking for:

$$K(x - y, -z, A)K(y, z, A') = \sum z^n H_n(x, y; A, A') \tag{3.3.1.1}$$

By construction, the integral

$$\int_{-\infty}^{\infty} H_n(x, y; A, A')\, dy = 0$$

for every $n > 0$. To get orthogonality of H_n with respect to all polynomials of degree less than n, consider

$$\sum z^n \int_{-\infty}^{\infty} y^k H_n(x, y; A, A')\, dy = \int_{-\infty}^{\infty} y^k K(x - y, -z, A)K(y, z, A')\, dy$$

where the terms of the summation must vanish for $k < n$. I.e., this must reduce to a polynomial in z of degree k. Or one can take the transform

$$\int_{-\infty}^{\infty} e^{sy}\, K(x - y, -z, A)K(y, z, A')\, dy$$

which has to be of the form such that the powers of z have factors depending on s so that each degree in z has a factor with a zero of at least that order in s, as observed in the remarks above.

3.3.2 Probabilities and means

Here is a general construction of kernels. Take any probability distributions whose means form an additive group. Suppose that they have densities. Then the kernels are of the form $K(x, z, A)$ where z is the mean, and A, e.g., is the variance, or other parameters determining the distribution. One example is provided by the Gaussian distributions:

$$K(x, z, A) = \frac{e^{-(x-z)^2/(2A)}}{\sqrt{2\pi A}}$$

Since means and variances are additive, you have a convolution family as required. In general, it may not always be possible to parametrize the family in terms of the means.

3.3.3 Bernoulli systems

Here we make a definition that applies for $N \geq 1$.

3.3.3.1 Definition. A *Bernoulli system* is a canonical Appell system such that the basis $\psi_n = R^n \Omega$ is orthogonal.

We have seen the example of the multinomial distributions. For $N = 1$, we have the binomial distributions, corresponding to Bernoulli trials, hence the name. We renormalize ψ_n and define a new generating function.

3.3.3.2 Definition. Define the basis $\phi_n = n! \times \psi_n / \gamma_n$, where $\gamma_n = \langle \psi_n, \psi_n \rangle$ are the squared norms of the ψ_n. The generating function ω^t is defined as

$$\omega^t(y, x) = \sum_{n \geq 0} \frac{y^n}{n!} \phi_n \tag{3.3.3.1}$$

Now we have an important property of ω^t.

3.3.3.3 Proposition. *Consider a Bernoulli system, in $N \geq 1$ dimensions, with canonical operator V and Hamiltonian H. I.e.,*

$$e^{z_\mu x_\mu - t H(z)} = \sum_{n \geq 0} \frac{V(z)^n}{n!} \psi_n$$

Let the basis ϕ_n and the function ω^t be as above. Then we have the Fourier-Laplace transform

$$\int e^{\zeta y} \omega^t(z, y) p_t(dy) = e^{z V(\zeta) + t H(\zeta)}$$

Proof: The integral on the left-hand side is the inner product

$$\langle e^{\zeta X}\,\Omega, \omega^t(z,X)\Omega\rangle = e^{tH(\zeta)}\,\langle e^{V(\zeta)R}\Omega, \omega^t(z,X)\Omega\rangle$$

By orthogonality, and the definition of ω^t, eq. (3.3.3.1), the inner product reduces to

$$\sum_{n\geq 0} \frac{z^n V(\zeta)^n n!\,\gamma_n}{n!\,n!\,\gamma_n} = e^{zV(\zeta)}$$

as required. ∎

Now go back to the case $N = 1$. Expanding in powers of z yields the relation

$$\int_{-\infty}^{\infty} e^{sy}\,\phi_n(y)\,p_t(dy) = V(s)^n\,e^{tH(s)}$$

so that $V(0) = 0$ is all we need to conclude that the ϕ_n are an orthogonal family. We take t as our parameter A and

$$K(x,z,A) = \omega^A(z,x)p_A(x) \tag{3.3.3.2}$$

writing $p_t(dx) = p_t(x)dx$ in the sense of distributions in the case of discrete spectrum (e.g., the Poisson case). In the case when $\omega^A(z,x) \geq 0$, these are a family of probability measures as noted in example 1, with mean $z + \mu t$, and variance $z + \sigma^2 t$, where μ and σ^2 are the mean and variance respectively of p_1.

We thus have from the basic construction, eq. (3.3.1.1),

$$K(x-y,-z,A)K(y,z,B) = \omega^A(-z,x-y)\omega^B(z,y)\,p_A(x-y)p_B(y) \tag{3.3.3.3}$$

Substituting in the expansions of the ω's, equation (3.3.3.1), yields

$$\sum_{n\geq 0} \frac{z^n}{n!} \sum_{k=0}^{n} \binom{n}{k}(-1)^k \phi_k(x-y,A)\phi_{n-k}(y,B)\,p_A(x-y)p_B(y)$$

Thus, the functions $H_n(x,y;A,B)$ take the form

$$H_n(x,y;A,B) = \sum_{k=0}^{n} \binom{n}{k}(-1)^k \phi_k(x-y,A)\phi_{n-k}(y,B)\,p_A(x-y)p_B(y)$$

with corresponding orthogonal polynomials

$$\phi_n(x,y;A,B) = \sum_{k=0}^{n} \binom{n}{k}(-1)^k \phi_k(x-y,A)\phi_{n-k}(y,B)$$

and measure of orthogonality $p_A(x - y)p_B(y)$. (Proof of orthogonality is based on the addition formula for $V(s)$, cf. Volume 1, Ch. 7, p. 199.) The convolution property of the family p_t shows that

$$\int_{-\infty}^{\infty} p_A(x - y)p_B(y)\,dy = p_{A+B}(x)$$

and thus, that we can normalize to give a probability measure of the form

$$p_A(x - y)p_B(y)/p_{A+B}(x)$$

For the Meixner classes, i.e., the Bernoulli systems in one variable corresponding to sl(2), we have the corresponding classes generated as follows:

Gaussian $\longrightarrow$ Gaussian
Poisson $\longrightarrow$ Krawtchouk
Laguerre $\longrightarrow$ Jacobi
Binomial (3 types) $\longrightarrow$ Hahn (3 types)

Observe that for the binomial types, this is essentially the construction of Clebsch-Gordan coefficients for (real forms of) sl(2). This construction works for the multinomial case of §3.2 as well.

3.3.4 Associativity construction

Corresponding to associativity of the convolution family, we form

$$K(x - y, -z, A + A')K(y, z, A''), \quad \text{and} \quad K(x - y, -z, A)K(y, z, A' + A'')$$

These both integrate to $K(x, 0, A + A' + A'')$. The corresponding $H_n(x, y; A + A', A'')$, $H_n(x, y; A, A' + A'')$ provide two orthogonal families for $L^2(dy)$. The question is to find the unitary transformation between the two bases, analogous to the construction of Racah coefficients.

For Bernoulli systems, denote the squared norms

$$\gamma_n(A, B) = \int_{-\infty}^{\infty} \phi_n(x, y; A, B)^2\, p_A(x - y)p_B(y)\,dy$$

Then we have the generating function for the unitary matrix U_{mn} connecting the combined systems corresponding to $A + B + C = (A + B) + C = A + (B + C)$, via equations (3.3.3.2), (3.3.3.3),

$$\int_{-\infty}^{\infty} \omega^A(-z, x - y)\omega^{B+C}(z, y)\omega^{A+B}(-w, x - y)\omega^C(w, y)$$

$$\times \sqrt{p_A(x - y)p_{B+C}(y)p_{A+B}(x - y)p_C(y)}\,dy$$

$$= \sum_{m,n} z^m w^n \sqrt{\gamma_m(A, B + C)}\, U_{mn}\, \sqrt{\gamma_n(A + B, C)}$$

For the binomial distributions, these will yield the usual Racah coefficients and connections with Wilson polynomials. Here we are just providing a sketch of the approach.

References

A useful reference on hypergeometric orthogonal polynomials and q-analogs is Koekoek & Swarttouw [41]. See Koornwinder & Sprinkhuizen-Kuyper [43] for an example of higher-dimensional Appell functions and hypergeometric functions on matrix spaces. Multivariate polynomials related to this chapter are considered in Tratnik [50].

Lorente [45] discusses construction of Hilbert spaces in a context similar to ours.

For coherent states and applications to physics, see: Hecht [32], Perelomov [47], Klauder & Skagerstam [39], and Zhang, Feng & Gilmore [57].

Chapters 6 and 7 of [19] are background for the present chapter. Also see [20] for properties of Bernoulli systems, with Krawtchouk polynomials considered in detail.

Chapter 5 Algebras with discrete spectrum

In this and the next three chapters, we will look at a variety of examples. Basic algebras include aff1, Heisenberg algebra (3-dimensional), finite-difference algebra, sl(2), e2, and a particular two-step nilpotent algebra that we call the 'ABCD' algebra (why — to be explained). And then some examples of higher-dimensional algebras.

I. Calculus on groups: review of the theory

First, let us summarize the theory we have developed.

1. A Lie algebra is given in terms of a typical element, $X = \alpha_\mu \xi_\mu$. As we will be using our MAPLE procedures, X will be a matrix. Thus, the basis $\{\xi_i\}$ is implicitly given. It is found explicitly by computing $\xi_i = \partial X/\partial \alpha_i$.

2. Given the ξ_i, one can directly compute the commutators and collect the results into a matrix, the Kirillov form. For Lie algebras with flags, the pi-matrices may be calculated using the adjoint representation.

3. Exponentiating and multiplying the one-parameter subgroups generated by the basis elements yields the group elements in terms of coordinates of the second kind, A. Multiplying the group elements, one can determine the group law.

4. The pi-matrices can be found via the splitting lemma, using the relation $\dot{g} = Xg = gX$. Thus, differential equations for the A's are found. Using the pi-matrices, the adjoint group can be computed. The adjoint representation may be quickly found via linearizing the difference between $\pi^* - \pi^{\ddagger}$.

5. From π^*, the right dual is found. Using the principal formula, the matrix elements for the action of the group on the universal enveloping algebra are given in recursive fashion. Using the double dual and the pi-matrices, one finds recurrence formulas for these matrix elements in terms of boson operators.

6. From the double dual, the multiplication rule for $\mathcal{U}(\mathcal{G})$ in terms of the action of the elements ξ_i on the basis for $\mathcal{U}(\mathcal{G})$ is given. Quotient representations can be determined using the double dual to find the action of $\mathcal{G}$ on quotients of $\mathcal{U}(\mathcal{G})$ by various ideals, equivalently, induced representations can be constructed.

7. By exponentiating $X \in \mathcal{G}$, one can look to construct a canonical Appell system. The factorization of $\mathbf{G}$ is essential; the splitting lemma is useful here. If one can find a Leibniz function, then it can be used to construct the representation, the Hilbert space and the evolution of a canonical Appell system.

8. Using the group law, one can find a set of integral equations for the Lie response: the interaction of the Lie structure with a signal. Stochastic processes on Euclidean

space yield in this way processes on the group. Wiener and Poisson processes are of particular interest here.

9. Appell systems corresponding to the given Lie structure can be constructed. Solutions can be given to 'heat equations' using the Lie response to the Wiener process. General analytic evolution equations are of interest as well.

1.1 MAPLE OUTPUT

Here follows a description of what the results on the MAPLE worksheet are in connection with the outline of the theory given above.

1. First the matrix $X = a_\mu \xi_\mu$ is given. If there is a choice, the basis is arranged so that it is orthogonal in the standard inner product: $\text{tr}(\xi_i \xi_j) = 0$ if $i \neq j$. However, for some algebras, a natural basis for the algebra may not have this property. The procedure grp, involving expanding in terms of the basis, is not run in such cases. The inputs to the procedure lie include $d = \dim \mathcal{G}$, n, where X is given as an $n \times n$ matrix, and 1 or 0, depending on whether the output is to be in trigonometric form or not. Details on the actual steps of the procedures are given in Chapter 9.

   First comes the group element $g(A)$, in terms of coordinates of the second kind. Then the exponential of the adjoint representation of the Lie algebra is given. Next the pi-matrices appear: $\pi^\dagger, \pi^*$.

2. The procedure grp gives the coefficients of the group element expanded in terms of the algebra basis $\{\xi_i\}$. These coefficients are useful for finding the group law, since they correspond, in part at least, to the expansion

$$e^{A\xi} = I + A\xi + \cdots$$

   In cases where $\xi^2 = 0$, and, in many realizations, many of the ξ are nilpotent, one can in fact directly read off corresponding coordinates for the group law.

3. The procedure matrec outputs a column array containing $\mathcal{R}_i$, raising the index n in $\left\langle \begin{smallmatrix} m \\ n \end{smallmatrix} \right\rangle$ by $\mathbf{e}_i$, in terms of boson operators. The operators RR and VV in the output indicate $\bar{\mathcal{R}}$ and $\bar{\mathcal{V}}$ acting on the index m. The operator $\mathcal{V}$ acts on n. Details on the underlying formula are in Chapter 8.

4. The procedure kirlv yields a matrix having i, j entry $c_{ij}^\mu x_\mu$, corresponding to the commutation relations $[\xi_i, \xi_j] = c_{ij}^\mu \xi_\mu$.

5. The duals are fairly self-explanatory. The operators are not necessarily ordered in the printed output. The partial derivative operators $\partial/\partial A_i$ are denoted δ_i. It is understood that all of the $\mathcal{V}$'s follow the $\mathcal{R}$'s, all δ's follow the A's. This is called "normal ordering."

6. The procedure indrep has an integer as input. This is typically $\dim \mathcal{P}$. The R_i become the variables of the representation, the raising operator for the next basis element of the Lie algebra is mapped to the scalar t, and all remaining variables R are set to

zero. For this to work, the basis must be appropriately ordered, as has been seen in Chapters 2 and 4. E.g., for the sl($N+1$) case, with the ordering R_i, ρ_0, ..., this gives the desired induced representation immediately in terms of boson operators.

II. Finite-difference algebra

Define the operators S, X, T on functions by

$$Sf(x) = \frac{f(x+h) - f(x)}{h}, \quad Xf(x) = xf(x), \quad Tf(x) = f(x+h)$$

where h is a fixed non-zero parameter, thought of as a discrete step-size or lattice spacing. They satisfy the commutation relations

$$[S,X] = T, \quad [T,X] = hT, \quad [S,T] = 0$$

We will refer to the Lie algebra spanned by these elements as the finite-difference algebra, the 'FD algebra' for short. Observe that the derived algebra $\mathcal{G}'$ is spanned by T, so is abelian, in particular nilpotent, so that the FD algebra is solvable, but, as is easily seen, not nilpotent. A matrix realization is given by

$$S = \frac{1}{\sqrt{2}} \begin{pmatrix} 1 & 1 \\ 0 & 1 \end{pmatrix}, X = \frac{1}{\sqrt{2}} \begin{pmatrix} -1 & 1 \\ 0 & 1 \end{pmatrix}, T = \begin{pmatrix} 0 & 1 \\ 0 & 0 \end{pmatrix}$$

with $h = \sqrt{2}$. We change basis, setting

$$P = (S - X)/\sqrt{2}, \quad Q = (S + X)/\sqrt{2}, \quad T = T$$

Thus, in matrix form,

$$Q = \begin{pmatrix} 0 & 1 \\ 0 & 1 \end{pmatrix}, T = \begin{pmatrix} 0 & 1 \\ 0 & 0 \end{pmatrix}, P = \begin{pmatrix} 1 & 0 \\ 0 & 0 \end{pmatrix}$$

It is clear from this realization that the Lie algebra is the same as the direct sum of the 2×2 diagonal matrices (abelian) with the upper-right corner.

From the matrix realization it is seen that $U = P + Q - T$ is in the center, as may be checked from the commutation relations as well. We thus take for our standard basis:

$$\xi_1 = Q, \quad \xi_2 = U, \quad \xi_3 = P$$

where, in the matrix realization, U is the identity matrix. The commutation relations read

$$[P,Q] = P + Q - U, \quad [P,U] = 0, \quad [U,Q] = 0$$

The Kirillov form is thus

$$\begin{pmatrix} 0 & 0 & -x_1 + x_2 - x_3 \\ 0 & 0 & 0 \\ x_1 - x_2 + x_3 & 0 & 0 \end{pmatrix}$$

An induced representation is given by $\xi_3\Omega = 0$, $\xi_2\Omega = t\Omega$, $\xi_1\Omega = R\Omega$. In terms of our matrix formulation, we can take $\Omega = \begin{pmatrix} 0 \\ 1 \end{pmatrix}$, with $t = 1$, as a specific example. We have the raising operator $R = \xi_1$, and we set $\rho = \xi_2$, $L = \xi_3$. It is natural to use for the group elements the form

$$g = (1 + V)^R e^{H\rho}(1 + B)^L$$

Denote the finite-difference operator with step size 1 on functions of R by δ_+: $\delta_+ f(R) = f(R+1) - f(R)$. Then δ_+ acts on $(1 + V)^R$ as multiplication by V. Another way to see why this is a useful change of variables is to observe that using the above matrices:

$$e^{sQ} = \begin{pmatrix} 1 & e^s - 1 \\ 0 & e^s \end{pmatrix} = \begin{pmatrix} 1 & V \\ 0 & 1 + V \end{pmatrix} \tag{2.1}$$

setting $e^s = 1 + V$. In matrix form, we have the group element

$$g = (1 + V)^R e^{H\rho}(1 + B)^L = \begin{pmatrix} e^H(1 + B) & Ve^H \\ 0 & (1 + V)e^H \end{pmatrix}$$

Thus, directly from this equation

2.1 Proposition.   *Given an element g of the FD group, we can recover the coordinates V, H, B by the relations:*

$$e^H = g_{22} - g_{12}, \quad V = g_{12}/e^H, \quad B = g_{11}/e^H - 1$$

Now

2.2 Theorem.   *The Leibniz rule is given by* $(1 + B)^L(1 + V)^R = g(\tilde{V}, \tilde{H}, \tilde{B})$ *where*

$$\tilde{V} = (1 + B)\frac{V}{1 - BV}, \quad e^{\tilde{H}} = 1 - BV, \quad \tilde{B} = B\frac{1 + V}{1 - BV}$$

Proof: By matrix multiplication, cf. eq. (2.1),

$$(1 + B)^L(1 + V)^R = \begin{pmatrix} 1 + B & (1 + B)V \\ 0 & 1 + V \end{pmatrix} = g(\tilde{V}, \tilde{H}, \tilde{B})$$

Using the previous Proposition, the result follows.   ∎

As the Leibniz rule shows that $e^{\tilde{H}}$ is symmetric in B, V, we can set up an inner product with R and L adjoints. First, we define the coherent states

$$\psi_V = (1 + V)^R \Omega$$

The inner product of the coherent states is thus

$$\Upsilon_{BV} = \langle \psi_B, \psi_V \rangle = (1 - BV)^t$$

for the induced representation with $\rho\Omega = t\Omega$. Now,

2.3 Proposition.   *The function* $\Upsilon_{BV} = (1 - BV)^t$ *satisfies*

$$(1 + B)\frac{\partial \Upsilon}{\partial B} = -tV\Upsilon + V(1 + V)\frac{\partial \Upsilon}{\partial V} \tag{2.2}$$

Proof: This can be checked directly.   ■

2.4 Proposition.   *In terms of functions of R, in the induced representation,*

$$L = (R - t)\delta_+$$

where we can express the difference operator $\delta_+ = e^{\partial_R} - 1$.

Proof: On the coherent state ψ_V, multiplication by R is given by $(1+V)\dfrac{\partial}{\partial V}$. Thus, for the action of L:

$$(1 + B)\frac{\partial \Upsilon}{\partial B} = \langle R\psi_B, \psi_V \rangle = \langle \psi_B, L\psi_V \rangle$$

Now, by the above Proposition, we have L according to the right side of eq. (2.2). Multiplication by V translates into the operator δ_+, and, as just remarked, $(1 + V)\partial/\partial V$ provides a factor of R.   ■

Observe that this is the same representation given as the induced representation derived from the double dual in the MAPLE output.

The coherent state $\psi_V = (1 + V)^R \Omega$ shows that the basis is given by

$$\psi_n = R(R - 1)\cdots(R - n + 1)\Omega \tag{2.3}$$

The operator δ_+ acts on ψ_n as the velocity operator, formal differentiation, $\mathcal{V}$. The operator R is not the raising operator here. In fact,

2.5 Proposition.   *A boson realization on the basis ψ_n is given by*

$$R = \mathcal{R} + \mathcal{R}\mathcal{V}, \qquad L = \mathcal{R}\mathcal{V} + (-t + \mathcal{R}\mathcal{V})\mathcal{V}$$

Proof: From eq. (2.3), we see that

$$R\psi_n = \psi_{n+1} + n\psi_n$$

hence the form of R in terms of $\mathcal{R}$ and $\mathcal{V}$. Now use the relation $L = (R - t)\delta_+$, and the result follows.   ■

Now we look for explicit realizations of the Hilbert space in terms of L^2 with respect to a probability measure. I.e., we want to see which elements of the Lie algebra can be interpreted as random variables. We calculate using matrices, writing $X = Q + \alpha P$:

$$e^{z(Q+\alpha P)} = \begin{pmatrix} e^{z\alpha} & (e^{z\alpha} - e^z)/(\alpha - 1) \\ 0 & e^z \end{pmatrix} \tag{2.4}$$

Therefore

2.6 Proposition. *The expected value of* $\exp\big(z(Q+\alpha P)\big)$ *is given by*

$$\langle e^{z(Q+\alpha P)}\,\rangle = \left(\frac{e^{z\alpha}-\alpha e^{z}}{1-\alpha}\right)^{t}$$

Proof: From eq. (2.4) and Proposition 2.1, follows that

$$e^{H} = e^{z} - \frac{e^{z\alpha}-e^{z}}{\alpha-1}$$

Raising to the power t and rearranging yields the result. ■

Now we have the following interesting feature of the FD algebra.

2.7 Theorem. *Let* $X = Q+\alpha P$ *be an element of the FD algebra. Then the distribution of* X *is as follows:*

1. *For* $\alpha < 0, t > 0$, *integer,* X *is distributed as the sum of independent Bernoulli random variables with jumps* $1, a$, *having probabilities* $a/(1+a), 1/(1+a)$, *respectively, where* $a = -\alpha$.

2. *For* $0 < \alpha < 1$, X *has a negative binomial distribution,*

$$P\left(X = n(1-\alpha)-\alpha\tau\right) = (1-\alpha)^{\tau}\,\alpha^{n}(\tau)_{n}/n!$$

where $t = -\tau$ *is any negative real number.*

3. *For* $\alpha = 1$, X *has a centered gamma distribution,*

$$\langle e^{zX}\,\rangle = e^{-z\tau}\,(1-z)^{-\tau}$$

for any negative real $t = -\tau$.

Proof: For #1, write $\alpha = -a$ in Proposition 2.6. to get

$$\frac{e^{az}}{1+a}+\frac{ae^{z}}{1+a}$$

raised to the power t. This is the moment generating function of a Bernoulli distribution as indicated. For #2, replace $t \to -\tau$ and factor out $(1-\alpha)^{\tau}e^{-z\alpha\tau}$ in Proposition 2.6. Expanding the remaining factor in powers of α shows the result. Finally, for #3, taking the limit $\alpha \to 1$ yields the indicated moment generating function. ■

Using these measures, we can explicitly construct the Hilbert spaces for the corresponding representations.

We conclude this section with the splitting formula for the FD group. Using eq. (2.4),

2.8 Proposition.   *For the FD group, we have the splitting*

$$e^{z(Q+\alpha P)} = \left(\frac{1-\alpha}{e^{(\alpha-1)z}-\alpha}\right)^Q \left(\frac{e^{z\alpha}-\alpha e^z}{1-\alpha}\right)^U \left(\frac{1-\alpha}{1-\alpha e^{(1-\alpha)z}}\right)^P$$

Proof: This follows from eq. (2.4) via Proposition 2.1.   ∎

It is interesting to compare this result with the splitting lemma, using the left and right dual matrices, $\pi^{\ddagger}$ and π^*, from the MAPLE output. We leave this to the reader.

III.   *q*-Heisenberg-Weyl algebra and basic hypergeometric functions

In the study of quantum groups, examples often involve 'q-analysis.' Here we provide a brief introduction to the area of q-analysis, but do not discuss quantum groups per se. Instead, we show how many of the constructions we have for groups carry over to the particular example of the q-Heisenberg case. From our perspective, the q-difference operator is a basic object, as the derivative corresponding to discrete, multiplicative shift by the factor q, analogous to the discrete additive shift in the finite-difference case.

Fix $0 < q < 1$. The q-derivative is given by

$$\delta_x f(x) = \frac{f(x) - f(xq)}{x - xq}$$

The q-Heisenberg algebra is the vector space with basis $\xi_1 = X$, multiplication by x, $\xi_2 = q^\nu$, where ν is the number operator, and $\xi_3 = \delta_x$. I.e., on the basis $[\![i]\!] = x^i$,

$$\xi_1[\![i]\!] = [\![i+1]\!], \quad \xi_2[\![i]\!] = q^i[\![i]\!], \quad \xi_3[\![i]\!] = q_i[\![i-1]\!]$$

where $q_i = \dfrac{1-q^i}{1-q}$ are the q-analogs of the usual integers. The ξ_i satisfy

$$[\xi_3,\xi_1] = \xi_2, \quad \xi_2\xi_1 = q\xi_1\xi_2, \quad \xi_3\xi_2 = q\xi_2\xi_3$$

i.e., in terms of the q-commutator, $[\xi,\eta]_q = \xi\eta - q\eta\xi$,

$$[\xi_2,\xi_1]_q = [\xi_3,\xi_2]_q = 0$$

We now recall some notations useful in this subject.

Notations.   These are commonly used, except the notation for the q-exponential varies somewhat among authors.

$$(x;q)_k = \prod_{j=0}^{k-1}(1-xq^j), \quad (x;q)_\infty = \prod_{j=0}^{\infty}(1-xq^j)$$

$$E(x) = \left((1-q)x;q\right)_\infty^{-1} = \sum_{k=0}^{\infty}\frac{x^k}{[k]!}$$

where the q-factorial $[k]! = q_k q_{k-1} \cdots q_1$. The q-binomial coefficients are denoted

$$\binom{j}{k}_q = \frac{(q;q)_j}{(q;q)_{j-k}(q;q)_k} = \frac{[j]!}{[j-k]![k]!}$$

The main feature of the q-exponential is that $E(ax)$ are eigenfunctions of δ_x:

$$\delta_x E(ax) = aE(ax)$$

as here we do not have a group composition law. One can define a composition in the space of functionals as follows. Think of $f(x)$ as $\varepsilon_x(f)$, where ε_x denotes the evaluation map at x. Denote the power function $x \to x^j$ by f_j. Then define the composition of ε_x, ε_y to be $\varepsilon_{x \oplus y}$ according to

$$\varepsilon_{x \oplus y}(f_j) = \sum_k \binom{j}{k}_q x^{j-k} y^k$$

So we write this informally as $(x \oplus y)^j$ and extend to power series. Then we have the relation

$$E(x \oplus y) = E(x)E(y)$$

meaning $\varepsilon_{x \oplus y}(E) = \varepsilon_x(E)\varepsilon_y(E)$. (The idea of working with the dual space connects up with the theory of quantum groups.)

The q-Heisenberg-Weyl algebra is the associative algebra with basis $[\![n]\!] = \xi_1^{n_1}\xi_2^{n_2}\xi_3^{n_3}$. The generating function of this basis is the analog of the group element:

$$g(A) = \sum_{n \geq 0} \frac{A^n}{[n]!} [\![n]\!] = E(A_1\xi_1)\,E(A_2\xi_2)\,E(A_3\xi_3)$$

where we use the multi-index notation $[n]! = [n_1]!\,[n_2]!\,[n_3]!$.

We compute the right dual directly, since there are no pi-matrices available. The following is shown by induction.

3.1 Proposition. *In the associative algebra, we have the commutation rules:*

1. $[\xi_3, \xi_1^j] = q_j \xi_1^{j-1} \xi_2$.
2. $[\xi_3^j, \xi_1] = \xi_2\, q_j \xi_3^{j-1}$.

And,

3.2 Corollary. *Exponential commutation rules:*

1. $\xi_3\, E(a\xi_1) = E(a\xi_1)(\xi_3 + a\xi_2)$.
2. $E(a\xi_3)\xi_1 = (\xi_1 + a\xi_2)E(a\xi_3)$.

Proof: Multiply by $a^n/[n]!$ in the above Proposition and sum.   ■

Now we want to find the Leibniz rule. First,

3.3 Lemma.   *q-binomial identity.   If $\eta\xi = q\xi\eta$, then*

$$(\xi + \eta)^j = \sum \binom{j}{k}_q \xi^{j-k}\eta^k$$

Proof: Shown by induction.   ■

And so,

3.4 Theorem.   *The Leibniz rule for q-Heisenberg-Weyl:*

$$E(B\xi_3)\,E(V\xi_1) = E(V\xi_1)E(BV\xi_2)E(B\xi_3)$$

Proof: From Corollary 3.2,

$$E(B\xi_3)\xi_1 E(B\xi_3)^{-1} = \xi_1 + B\xi_2$$

Now, by Lemma 3.3,

$$E(B\xi_3)\xi_1^j E(B\xi_3)^{-1} = \sum_k \binom{j}{k}_q \xi_1^{j-k}(B\xi_2)^k$$

Multiplying by $V^j/[j]!$ and summing yields the result.   ■

We now construct the q-HW Hilbert space. Set $R = \xi_1$, $\rho = \xi_2$, $L = \xi_3$. The induced representation is given by the conditions $L\Omega = 0$, $\rho\Omega = \Omega$. The coherent states have the form

$$\psi_V = E(VR)\Omega$$

With R and L adjoints, the Leibniz function is

$$\langle \psi_B, \psi_V \rangle = \langle \Omega, E(BL)E(VR)\Omega \rangle \\ = E(BV) \tag{3.1}$$

via Theorem 3.4. The cocycle identity has the form $\Upsilon_{B,U\oplus V} = \Upsilon_{UB}\Upsilon_{BV}$.

3.5 Proposition.   *The basis $\psi_j = R^j\Omega$ is orthogonal, with squared norms $\gamma_j = [j]!$.*

Proof: Expand both sides of the first and last expressions in eq. (3.1). Then comparing the expansions

$$E(BV) = \sum \frac{B^j V^j}{[j]! \, [j]!} \gamma_j$$

yields the result. ∎

Next, we find the right dual.

3.6 Theorem. *The right dual, acting by the basis elements on $E(A_1\xi_1)E(A_2\xi_2)E(A_3\xi_3)$ is*

$$\xi_1^* = q^{\nu_2}\delta_1 + A_3\delta_2 , \quad \xi_2^* = q^{\nu_3}\delta_2 , \quad \xi_3^* = \delta_3$$

where δ_i denotes the q-derivative with respect to A_i, and q^{ν_i} scales $A_i \to qA_i$.

Proof: For ξ_2, use the q-commutation relation

$$E(A_3\xi_3)\xi_2 = \xi_2 E(qA_3\xi_3)$$

For ξ_1, use Corollary 3.2. ∎

Here we can find the matrix elements for the representation of the action of the group-like $g(A)$ on the q-Heisenberg-Weyl algebra. The principal formula, Ch.2, Theorem 3.1.1, applies here, since we have a well-defined right dual. The difference is that now the coefficients $c_m(A) = A^m/[m]!$. Recall the *basic hypergeometric function* denoted $_2\phi_1$:

$$_2\phi_1 \left(\begin{array}{c} a, b \\ c \end{array} \middle| x \right) = \sum \frac{(a;q)_k (b;q)_k}{(c;q)_k (q;q)_k} x^k$$

Then

3.7 Theorem. *The matrix elements for the action of $g(A)$ on the q-Heisenberg-Weyl algebra are*

$$\left\langle \begin{array}{c} m \\ n \end{array} \right\rangle_A = \frac{A^\Delta}{[\Delta]!} q^{n_1\Delta_2 + n_2\Delta_3} \, _2\phi_1 \left(\begin{array}{c} q^{-n_1}, q^{-\Delta_2} \\ q^{1+\Delta_1} \end{array} \middle| \frac{qA_1 A_3}{A_2} \right)$$

where $\Delta = m - n$.

Remark. Note that we are only considering the case $\Delta \geq 0$.

Proof: The principal formula yields

$$(\xi^*)^n c_m(A) = (q^{\nu_2}\delta_1 + A_3\delta_2)^{n_1} (q^{\nu_3}\delta_2)^{n_2} \delta_3^{n_3} \frac{A^m}{[m]!}$$

Using Lemma 3.3, writing $\Delta_i = m_i - n_i$,

$$\left\langle \begin{array}{c} m \\ n \end{array} \right\rangle_A = \binom{n_1}{\lambda}_q (q^{\nu_2}\delta_1)^{n_1-\lambda} (A_3\delta_2)^\lambda \frac{A_1^{m_1}}{[m_1]!} \frac{A_2^{\Delta_2}}{[\Delta_2]!} \frac{(q^{n_2}A_3)^{\Delta_3}}{[\Delta_3]!}$$

$$= \binom{n_1}{\lambda}_q \frac{A_1^{\Delta_1+\lambda}}{[\Delta_1+\lambda]!} \frac{(q^{n_1-\lambda}A_2)^{\Delta_2-\lambda}}{[\Delta_2-\lambda]!} \frac{q^{n_2\Delta_3} A_3^{\Delta_3+\lambda}}{[\Delta_3]!}$$

which simplifies down to

$$q^{n_2\Delta_3+n_1\Delta_2}\frac{A^\Delta}{[\Delta]!}\frac{(q^{-n_1};q)_\lambda(q^{-\Delta_2};q)_\lambda}{(q^{1+\Delta_1};q)_\lambda(q;q)_\lambda}\left(\frac{qA_1A_3}{A_2}\right)^\lambda$$

which is now written as a $_2\phi_1$. ■

IV. su2 and Krawtchouk polynomials

The Krawtchouk polynomials arise as the orthogonal polynomials with respect to the binomial distribution. The Krawtchouk matrices Φ defined below have many important properties in probability theory and combinatorics as well. In probability theory, the Krawtchouk polynomials, evaluated along the paths of a symmetric random walk, with steps ±1, are martingales in the parameter N, the 'time.' The combinatorial feature of principal interest is that each column of the matrices Φ satisfies Pascal's triangle. Thus, in a very direct way the Φ are generalizations of binomial coefficients.

Other interpretations revolve around the underlying ±1's considered as a discrete field of spins. From this point of view, the Φ's and related mathematics provide a basic discrete model of quantum mechanics.

4.1 Definition. The *Krawtchouk matrices* are defined through the generating function

$$(1+v)^{N-j}(1-v)^j = \sum_{k=0}^{N}v^k\Phi_{kj} \tag{4.1}$$

with $\Phi_{kj} = \phi_k(j)$ the kth *Krawtchouk polynomial.*

This generating function is the simplest example of that which we found for the multinomial case, Chapter 4, Theorem 3.2.6.

Remark. Note that the matrices appearing throughout this section depend on the index N. This dependence is implicit throughout the discussion.

WARNING: In this section N denotes the 'time' in the underlying random walk: a discrete time parameter. It is the time-parameter t of the general discussion.

4.1 FINITE-DIMENSIONAL REPRESENTATIONS OF sl(2)

Consider the Lie algebra with basis $\{R,\Lambda,L\}$ satisfying the commutation relations

$$[L,R]=\Lambda,\qquad[R,\Lambda]=2R,\qquad[\Lambda,L]=2L$$

The identification

$$R = \begin{pmatrix} 0 & 0 \\ 1 & 0 \end{pmatrix}, \quad L = \begin{pmatrix} 0 & 1 \\ 0 & 0 \end{pmatrix}, \quad \Lambda = \begin{pmatrix} 1 & 0 \\ 0 & -1 \end{pmatrix}$$

shows this to be the Lie algebra, sl(2), of two by two matrices of trace zero. Another convenient basis is given by

$$X = R + L, \quad \Lambda, \quad Y = R - L$$

satisfying

$$[X, Y] = 2\Lambda \quad [Y, \Lambda] = 2X, \quad [X, \Lambda] = 2Y \qquad (4.1.1)$$

Remark. To see how this fits in with su(2), we note that the connection with angular momentum theory is given by

$$J_1 = X, \quad J_2 = iY, \quad J_3 = \Lambda$$

For $N = 1$ these are the Pauli matrices.

For general N,

$$R = \begin{pmatrix} 0 & 0 & 0 & \cdots \\ N & 0 & 0 & \cdots \\ 0 & N-1 & 0 & \cdots \\ 0 & 0 & N-2 & 0 & \cdots \\ \vdots & \vdots & \vdots & \ddots & \ddots & \vdots \\ & & & & 1 & 0 \end{pmatrix}$$

$$L = \begin{pmatrix} 0 & 1 & 0 \cdots \\ 0 & 0 & 2 \cdots \\ 0 & 0 & 0 & 3 \cdots \\ \vdots & \vdots & \vdots & \ddots & \ddots & \vdots \\ 0 \cdots & & & & N \\ 0 \cdots & & & & 0 \end{pmatrix} \qquad (4.1.2)$$

and

$$\Lambda = [L, R] = \begin{pmatrix} N & 0 & \cdots & & 0 \\ 0 & N-2 & 0 & \cdots \\ \vdots & \vdots & \ddots & & \vdots & \vdots \\ 0 \cdots & & & 2-N & 0 \\ 0 \cdots & & & & -N \end{pmatrix}$$

Remark. Note that matrices are indexed from 0 to N.

These give the finite-dimensional irreducible representations of sl(2,**R**) (see Satake [48], pp. 89–90). A useful observation is the following

4.1.1 Proposition.   *In any representation,*

$$R\Lambda = (\Lambda + 2)R$$
$$R\,a^\Lambda = a^\Lambda\,(a^2 R)$$
$$e^R\,a^\Lambda = a^\Lambda\,e^{a^2 R}$$

for $a > 0$, with similar equations holding for L.

The principal group properties are summarized in the following two propositions. We refer to Volume 1 for the details. We remark that the idea of proving these relations is to differentiate one-parameter subgroups with respect to parameters and use the commutation rules to verify agreement. (The parameters are real numbers.)

4.1.2 Proposition.   *As a variation on the Leibniz rule,*

$$e^{tL}\,e^{aR} = (1 + at)^{\Lambda/2}e^{aR}\,e^{tL}\,(1 + at)^{\Lambda/2}$$

for $|at| < 1$.

Proof: Referring to Volume 1, Chapter 1, Proposition 3.1.1, use the Lie algebra isomorphism:

$$L \to -\Delta,\quad R \to R,\quad \Lambda \to -\rho$$

From Proposition 4.1.1, we see that the result is equivalent to the Leibniz rule

$$e^{tL}\,e^{aR} = \exp\left(\frac{a}{1 + at}\,R\right)(1 + at)^\Lambda \exp\left(\frac{t}{1 + at}\,L\right)$$

Now apply Proposition 3.3.2 of Volume 1, Ch.1 .   ∎

One can extend the parameters to the complex plane. Then by analytic continuation, the formula holds for any branch of $\log(1 + at)$, for example, for at in the right half plane. Taking $a = t = 1$ we have the result

4.1.3 Corollary.   *The commutation rule*

$$e^L\,e^R\,2^{-\Lambda/2} = 2^{\Lambda/2}e^R\,e^L$$

which in matrix form can be expressed in terms of binomial coefficients (see below, Proposition 4.1.7.)

4.1.4 Proposition.   *The splitting formula*

$$e^{sR+u\Lambda+tL} = \exp\left(\frac{sT}{\delta + uT}R\right)\left((1 + \frac{u}{\delta}T)\cosh\delta\right)^\Lambda \exp\left(\frac{tT}{\delta + uT}L\right)$$

where $\delta^2 = u^2 + st$, $T = \tanh\delta$.

Proof: As in the above proof, apply Proposition 4.3.1 of Volume 1, Ch.1 .   ∎

With $Y = R - L$,

4.1.5 Corollary. *The exponential formula for Y:*

$$e^{sY} = e^{(\tan s)R} (\cos s)^{\Lambda} e^{(-\tan s)L}$$

for $|s| < \frac{\pi}{2}$.

Next we have the connection with the matrix of binomial coefficients.

4.1.6 Definition. The matrix with entries binomial coefficients is denoted β. That is,

$$\beta_{ij} = \binom{i}{j}$$

The main feature is

4.1.7 Proposition. *The exponential of L in the realization given above, eq. (4.1.2), is:*

$$e^{L} = \beta^{\dagger}$$

in fact, the exponential e^{tL} has entries $\binom{j}{i} t^{j-i}$.

Proof: Use the fact that L acts on a basis e_k as $Le_k = ke_{k-1}$.   ∎

From the group-theoretical point of view, one can define β in terms of L via this relation.

4.1.1 Binomial coefficient matrix and transposition symmetry

4.1.1.1 Definition. Denote by $\mathcal{B}$ the diagonal matrix with entries $\mathcal{B}_{ii} = \binom{N}{i}$.

Checking on a basis in the matrix realization, we see

4.1.1.2 Proposition. *As an operator on the Lie algebra, $\mathcal{B}$ acts as a transposition symmetry in the following manner:*

$$R\mathcal{B} = \mathcal{B}L^{\dagger} \qquad \text{and} \qquad L\mathcal{B} = \mathcal{B}R^{\dagger}$$

and

$$\mathcal{B}X^{\dagger} = X\mathcal{B} \qquad \text{and} \qquad \mathcal{B}Y^{\dagger} = -Y\mathcal{B}$$

4.2 REFLECTIONS

There are two operators that play an important role in the symmetry of the Krawtchouk polynomial systems.

4.2.1 Definition. The *parity operator* is given by

$$\sigma = (-1)^{(N-\Lambda)/2}$$

And

4.2.2 Definition. The *inversion operator* is

$$J = e^{-L} e^{R} e^{-L} \sigma$$

Note that this can be written as

$$e^{-L} e^{R} \sigma e^{L}$$

It is easy to see that $e^{R} \sigma$ is a reflection, since $\sigma R = -R\sigma$.

4.2.3 Proposition. *In matrix terms with R, Λ, L given above we have*

1. *The parity operator is diagonal with entries $\sigma_{ii} = (-1)^{i}$, $0 \leq i \leq N$.*

2. *The inversion operator has entries $J_{ij} = 1$, for $i + j = N$, 0 otherwise.*

Both of these are reflections: $\sigma^2 = J^2 = I$.

We note that

4.2.4 Proposition. *Commutation relations with σ:*

$$\sigma R = -R\sigma, \qquad \sigma L = -L\sigma$$

And for J,

$$R = JLJ$$

Proof: For σ use Proposition 4.1.1, that Λ is shifted by 2 on commuting past R or L. For J, we see that the relation is equivalent to the equality $e^{L} R e^{-L} = -e^{R} L e^{-R}$. These have the common form $R + \Lambda - L$. E.g., the adjoint action:

$$e^{L} R e^{-L} = e^{ad\, L} R$$

is computed using the commutation relations:

$$e^{L} R e^{-L} = R + [L, R] + \tfrac{1}{2}[L, [L, R]] + \cdots$$

with third-order commutators and corresponding higher-order terms identically zero. A similar computation with $e^{R} L e^{-R}$ gives the result. ■

Similar relations hold for X and Y, while both σ and J commute with Λ.

Combining the above with Proposition 4.1.7 yields

4.2.5 Corollary. β *satisfies*

$$\beta^{-1} = \sigma\beta\sigma$$

This result in the matrix realization of Definition 4.1.6 is, of course, well-known.

We define

4.2.6 Definition. Krawtchouk matrices with parameter s

$$\tilde{\Phi}(s) = e^{sY}\,\sigma$$

And

4.2.7 Lemma. *In the matrix realization of the group, we have the factorization of* Φ

$$\Phi = e^R\,2^{(N-\Lambda)/2}e^{-L}\,\sigma$$

where we recall that $(N-\Lambda)/2$ *is diagonal with entries* $0, 1, 2, \ldots, N$.

Proof: Write the generating function for Φ

$$(1+v)^{N-j}(1-v)^j = (1+v)^{N-j}(1+v-2v)^j$$

$$= (1+v)^N\left(1 - \frac{2v}{1+v}\right)^j$$

$$= \sum_\varepsilon \binom{j}{\varepsilon}(-2v)^\varepsilon(1+v)^{N-\varepsilon}$$

Rewrite this as

$$v^N\sum_\varepsilon \binom{j}{\varepsilon}(-2)^\varepsilon(1+v^{-1})^{N-\varepsilon} = \sum_{\varepsilon,k}\binom{j}{\varepsilon}(-2)^\varepsilon\binom{N-\varepsilon}{k}v^{N-k}$$

Thus,

$$\sum_k v^k\Phi_{kj} = \sum_k v^k\sum_\varepsilon \binom{N-\varepsilon}{N-k}\binom{j}{\varepsilon}(-2)^\varepsilon$$

This says, in matrix form,

$$\Phi = J\beta^\dagger J(-2)^{(N-\Lambda)/2}\beta^\dagger$$

and the result follows from the definition of σ and Propositions 4.1.7 and 4.2.4.   ∎

4.2.8 Theorem. *In terms of the group,* Φ *can be expressed in the form*

$$\Phi = 2^{N/2}\,e^{(\pi/4)Y}\,\sigma$$

Proof: This follows from Corollary 4.1.5 and the above Lemma.   ■

From the commutation properties of σ ·

4.2.9 Proposition.   *For all s,*

$$\tilde{\Phi}(s) = e^{(s/2)Y}\,\sigma e^{-(s/2)Y}$$

4.2.10 Corollary.   *For all s, $\tilde{\Phi}(s)$ is similar to σ. Hence, the spectrum of $\tilde{\Phi}(s)$ is the same as that of σ.*

4.2.11 Corollary. Φ *and Y anticommute:*

$$\Phi Y + Y\Phi = 0$$

4.2.12 Corollary. $\tilde{\Phi}(s)^2 = I$. *Hence,* $\Phi^2 = 2^N I$.

Regarding Φ, σ, and J,

4.2.13 Proposition.   *We have:*

$$J\Phi = \Phi\sigma$$
$$\Phi J = \sigma\Phi$$

Proof: Note that the second line follows from the first by taking inverses of both sides. The first line follows from Lemma 4.2.7 and Corollary 4.1.3 as we will now see. Write $\tilde{\Phi}$ for $2^{-N/2}\Phi$. Then, from the definition of J

$$
\begin{aligned}
J\tilde{\Phi} &= e^{-L}\,e^R\,\sigma \cdot e^L\,e^R\,2^{-\Lambda/2} \cdot e^{-L}\,\sigma \\
&= e^{-L}\,e^R\,\sigma \cdot 2^{\Lambda/2}e^R\,e^L \cdot e^{-L}\,\sigma \\
&= e^{-L}\,e^R\,2^{\Lambda/2}e^{-R}\,\sigma \cdot \sigma
\end{aligned}
$$

using Corollary 4.1.3 for the second line. So, by Proposition 4.1.1, dropping factors of σ, we must show that

$$e^{-L}\,2^{\Lambda/2}e^R = e^R\,2^{-\Lambda/2}e^{-L}$$

which again is Corollary 4.1.3.   ■

4.3 ROTATIONS AND RECURRENCE RELATIONS

Relations among Φ, X, and Λ have a group-theoretic basis in the adjoint action of the one-parameter subgroup $\exp(sY)$.

4.3.1 Lemma.   On the pair $\begin{pmatrix} X \\ \Lambda \end{pmatrix}$, the adjoint action of e^{sY}, i.e., conjugation by e^{sY}, is rotation by $2s$.

Proof: We will show that

$$F(s) = e^{sY} \begin{pmatrix} X \\ \Lambda \end{pmatrix} e^{-sY} = e^{s\,\mathrm{ad}\,Y} \begin{pmatrix} X \\ \Lambda \end{pmatrix}$$

satisfies

$$F(0) = \begin{pmatrix} X \\ \Lambda \end{pmatrix}, \quad F'(0) = 2 \begin{pmatrix} -\Lambda \\ X \end{pmatrix}, \quad \text{and } F'' + 4F = 0$$

which will make the identification

$$F(s) = \begin{pmatrix} \cos 2s & -\sin 2s \\ \sin 2s & \cos 2s \end{pmatrix} \begin{pmatrix} X \\ \Lambda \end{pmatrix}$$

By the commutation relations, eq. (4.1.1),

$$F'(s) = e^{s\,\mathrm{ad}\,Y} (\mathrm{ad}\,Y) \begin{pmatrix} X \\ \Lambda \end{pmatrix} = e^{s\,\mathrm{ad}\,Y} \begin{pmatrix} -2\Lambda \\ 2X \end{pmatrix}$$

Differentiating once again and using the commutation relations gives the result. ∎

4.3.2 Lemma.   The adjoint action of $\tilde{\Phi}(s)$ on $\begin{pmatrix} X \\ \Lambda \end{pmatrix}$ is given by

$$\tilde{\Phi}(s) \begin{pmatrix} X \\ \Lambda \end{pmatrix} = \begin{pmatrix} -\cos 2s & \sin 2s \\ \sin 2s & \cos 2s \end{pmatrix} \begin{pmatrix} X \\ \Lambda \end{pmatrix} \tilde{\Phi}(s)$$

Proof: The additional factor of σ acts as

$$\sigma \begin{pmatrix} X \\ \Lambda \end{pmatrix} = \begin{pmatrix} -X \\ \Lambda \end{pmatrix} \sigma$$

and then the above Proposition applies. ∎

4.3.3 Theorem. Φ satisfies

$$\Phi \begin{pmatrix} X \\ \Lambda \end{pmatrix} = \begin{pmatrix} \Lambda \\ X \end{pmatrix} \Phi$$

Proof: In the Lemma, evaluate at $s = \pi/4$.   ■

The equations

$$X\Phi = \Phi\Lambda$$
$$\Phi X = \Lambda\Phi$$

say that the columns of Φ are the eigenvectors of X with spectrum Λ and dually for the rows. Since X is tridiagonal and Λ is diagonal, we see that these are in fact the *recurrence relations* and *dual recurrence relations* for the orthogonal polynomials $\phi_k(j)$, the Krawtchouk polynomials.

The MAPLE output for sl(2) and so(3) has been included. Note that the matrix recurrences have been rescaled for printing.

For sl(2): a factor of $e^{2(A_2+V_2)}$ has been put in.

For so(3): a factor of $\cos(\bar{\mathcal{R}}_2)$, i.e., of $\cos(RR_2)$, has been put in.

V.   e2 and Lommel polynomials

The Euclidean group in two dimensions is the semidirect product of the rotation group — the circle — and the translations. Acting on functions $f(x,y)$, we take

$$\xi_1 = \frac{\partial}{\partial x}, \quad \xi_2 = \frac{\partial}{\partial y}, \quad \xi_3 = x\frac{\partial}{\partial y} - y\frac{\partial}{\partial x}$$

as basis for the Lie algebra. They satisfy the commutation relations

$$[\xi_1, \xi_2] = 0, \quad [\xi_1, \xi_3] = \xi_2, \quad [\xi_3, \xi_2] = \xi_1$$

The adjoint representation is given by

$$\xi_1 = \begin{pmatrix} 0 & 0 & 0 \\ 0 & 0 & 1 \\ 0 & 0 & 0 \end{pmatrix}, \quad \xi_2 = \begin{pmatrix} 0 & 0 & -1 \\ 0 & 0 & 0 \\ 0 & 0 & 0 \end{pmatrix}, \quad \xi_3 = \begin{pmatrix} 0 & 1 & 0 \\ -1 & 0 & 0 \\ 0 & 0 & 0 \end{pmatrix}$$

This is a faithful representation.

It is convenient to use complex coordinates. Setting $Z = \xi_1 + i\xi_2$, $\bar{Z} = \xi_1 - i\xi_2$, $\Theta = \xi_3$, we have the matrices

$$Z = \begin{pmatrix} 0 & 0 & -i \\ 0 & 0 & 1 \\ 0 & 0 & 0 \end{pmatrix}, \quad \bar{Z} = \begin{pmatrix} 0 & 0 & i \\ 0 & 0 & 1 \\ 0 & 0 & 0 \end{pmatrix}, \quad \Theta = \begin{pmatrix} 0 & 1 & 0 \\ -1 & 0 & 0 \\ 0 & 0 & 0 \end{pmatrix}$$

and the commutation relations

$$[Z, \bar{Z}] = 0, \quad [\Theta, Z] = iZ, \quad [\Theta, \bar{Z}] = -i\bar{Z}$$

These last two become in the enveloping algebra the translations:

$$\Theta Z = Z(\Theta + i), \quad \Theta\bar{Z} = \bar{Z}(\Theta - i)$$

Observe that we have a Cartan decomposition, with $\mathcal{P}$ and $\mathcal{L}$ spanned by Z and $\bar{Z}$ respectively, and $\mathcal{K}$ spanned by Θ. In this case **K**, the circle group, is indeed compact (the origin of the notation). From this action follows for corresponding one-parameter subgroups

5.1 Proposition. *Commutation properties*

$$e^{t\Theta} e^{sZ} = e^{se^{it}Z} e^{t\Theta}$$

$$e^{s\bar{Z}} e^{t\Theta} = e^{t\Theta} e^{se^{it}\bar{Z}}$$

From the action of Θ on Z and $\bar{Z}$ we see that the *Casimir operator*, a second-degree element in the center of $\mathcal{U}(\mathcal{G})$, is $Z\bar{Z}$. Thus, for irreducible representations, we will study representations with the property

$$Z\bar{Z} = \lambda$$

for some scalar λ. Note that we cannot then have $\bar{Z}$ acting as zero on the vacuum state. In this case, Ω more properly acts as a cyclic vector as will be seen. This is why we do not have an indrep procedure run for this algebra in MAPLE. We will find the splitting formula and the dual representations for comparison with the MAPLE output.

NOTE: Throughout this section m, n denote single indices.

5.1 GROUP LAW

Define group elements

$$g(A_1, A_2, A_3) = e^{A_1 Z} e^{A_2 \Theta} e^{A_3 \bar{Z}}$$

In terms of the above matrices one readily finds

5.1.1 Proposition. *We have the matrices*

$$e^{A_1 Z} = \begin{pmatrix} 1 & 0 & -iA_1 \\ 0 & 1 & A_1 \\ 0 & 0 & 1 \end{pmatrix}, \; e^{A_2 \Theta} = \begin{pmatrix} \cos A_2 & \sin A_2 & 0 \\ -\sin A_2 & \cos A_2 & 0 \\ 0 & 0 & 1 \end{pmatrix}, \; e^{A_3 \bar{Z}} = \begin{pmatrix} 1 & 0 & iA_3 \\ 0 & 1 & A_3 \\ 0 & 0 & 1 \end{pmatrix}$$

and the group elements of the form

$$g(A_1, A_2, A_3) = \begin{pmatrix} \cos A_2 & \sin A_2 & i(A_3 e^{-iA_2} - A_1) \\ -\sin A_2 & \cos A_2 & A_3 e^{-iA_2} + A_1 \\ 0 & 0 & 1 \end{pmatrix}$$

Multiplying these together, we have the group law

$$g(A_1, A_2, A_3)g(A_1', A_2', A_3') = g(A_1 + A_1' e^{iA_2}, A_2 + A_2', A_3 e^{iA_2'} + A_3')$$

which can be verified using the commutation relations, Prop. 5.1.

5.2 DUAL REPRESENTATIONS

Now we use the Splitting Lemma. Write the one-parameter subgroup generated by X in the form

$$e^{sX} = g(A_1(s), A_2(s), A_3(s))$$

Differentiating with respect to s yields

$$Xg = \dot{g}$$

from which we can read off the pi-matrices, and find the dual representations. Solving the equations gives the coordinates of the second kind corresponding to factorization of the group elements. First,

5.2.1 Proposition. For $X = \alpha_1 Z + \alpha_2 \Theta + \alpha_3 \bar{Z}$, the coordinates $(A_1(s), A_2(s), A_3(s))$ satisfy

$$\dot{A}_1 = i\alpha_2 A_1 + \alpha_1, \quad \dot{A}_2 = \alpha_2, \quad \dot{A}_3 = \alpha_3 e^{iA_2}$$

according to the left action of multiplication by X.

Proof: As matrices,

$$X = \begin{pmatrix} 0 & \alpha_2 & i(\alpha_3 - \alpha_1) \\ -\alpha_2 & 0 & \alpha_3 + \alpha_1 \\ 0 & 0 & 0 \end{pmatrix}$$

and

$$\dot{g} = \begin{pmatrix} -\dot{A}_2 \sin A_2 & \dot{A}_2 \cos A_2 & (i\dot{A}_3 + \dot{A}_2 A_3)e^{-iA_2} - i\dot{A}_1 \\ -\dot{A}_2 \cos A_2 & -\dot{A}_2 \sin A_2 & (\dot{A}_3 - i\dot{A}_2 A_3)e^{-iA_2} + \dot{A}_1 \\ 0 & 0 & 0 \end{pmatrix}$$

Multiplying out $Xg(s)$ and comparing with $\dot{g}$, the result follows. ∎

Now set up the differential equations, using this Proposition:

$$\begin{pmatrix} A_1 & A_2 & A_3 \end{pmatrix}^{\bullet} = \begin{pmatrix} \alpha_1 & \alpha_2 & \alpha_3 \end{pmatrix} \begin{pmatrix} 1 & 0 & 0 \\ iA_1 & 1 & 0 \\ 0 & 0 & e^{iA_2} \end{pmatrix}$$

Then we find the left dual via

$$\begin{pmatrix} Z^{\ddagger} & \Theta^{\ddagger} & \bar{Z}^{\ddagger} \end{pmatrix} = \begin{pmatrix} 1 & 0 & 0 \\ iA_1 & 1 & 0 \\ 0 & 0 & e^{iA_2} \end{pmatrix} \begin{pmatrix} \partial/\partial A_1 \\ \partial/\partial A_2 \\ \partial/\partial A_3 \end{pmatrix}$$

That is, we find the left dual representation

$$Z^{\ddagger} = \partial_1$$
$$\Theta^{\ddagger} = \partial_2 + iA_1 \partial_1$$
$$\bar{Z}^{\ddagger} = e^{iA_2} \partial_3$$

and the double dual

$$\hat{Z} = \mathcal{R}_1$$
$$\hat{\Theta} = \mathcal{R}_2 + i\mathcal{R}_1 \mathcal{V}_1$$
$$\hat{\bar{Z}} = \mathcal{R}_3 e^{i\mathcal{V}_2}$$

Similarly, multiplication on the right yields the system

$$(A_1 \quad A_2 \quad A_3)^* = (\alpha_1 \quad \alpha_2 \quad \alpha_3) \begin{pmatrix} e^{iA_2} & 0 & 0 \\ 0 & 1 & iA_3 \\ 0 & 0 & 1 \end{pmatrix}$$

with the resulting right dual representation

$$Z^* = e^{iA_2} \partial_1$$
$$\Theta^* = \partial_2 + iA_3\partial_3$$
$$\bar{Z}^* = \partial_3$$

One feature of using matrices is that the commutation rules do not have to be known *a priori*.

5.3 SPLITTING FORMULA

Solving the differential equations of Prop. 5.2.1 yields

5.3.1 Proposition. *Splitting formula*

$$e^{\alpha_1 Z + \alpha_2 \Theta + \alpha_3 \bar{Z}} = g\left(\frac{\alpha_1}{i\alpha_2}(e^{i\alpha_2} - 1), \alpha_2, \frac{\alpha_3}{i\alpha_2}(e^{i\alpha_2} - 1) \right)$$

In particular,

5.3.2 Proposition. *Flow for* $X = Z + \varepsilon\Theta + \bar{Z}$:

$$e^{sX} = g\left(\frac{1}{i\varepsilon}(e^{is\varepsilon} - 1), s\varepsilon, \frac{1}{i\varepsilon}(e^{is\varepsilon} - 1) \right)$$

Replacing ε by $-i\varepsilon$

5.3.3 Proposition. *If* $X = Z - i\varepsilon\Theta + \bar{Z}$, *then*

$$e^{sX} = g\left(\frac{1}{\varepsilon}(e^{s\varepsilon} - 1), -is\varepsilon, \frac{1}{\varepsilon}(e^{s\varepsilon} - 1) \right)$$

We can write this in the form

$$e^{sX} = e^{\frac{1}{\varepsilon}(e^{s\varepsilon} - 1)Z} e^{\frac{1}{\varepsilon}(1 - e^{-s\varepsilon})\bar{Z}} e^{-is\varepsilon\Theta} \tag{5.3.1}$$

And substituting $\xi = e^{\varepsilon s}$ gives

$$\xi^{X/\varepsilon} = e^{(\xi - 1)Z/\varepsilon} e^{(1 - 1/\xi)\bar{Z}/\varepsilon} \xi^{-i\Theta}$$

We want an element of the Lie algebra X that will act as a self-adjoint operator on a Hilbert space and thus will give us a random variable. Here we take

$$X = Z - i\varepsilon\Theta + \bar{Z}$$

The Lie algebra acts thus:

5.3.4 Proposition. *Action on functions of X:*

$$Zf(X) = f(X - \varepsilon)Z, \qquad \bar{Z}f(X) = f(X + \varepsilon)\bar{Z}$$

with $\Theta = i(X - Z - \bar{Z})/\varepsilon$.

5.4 REPRESENTATIONS

From the double dual, for $X = Z - i\varepsilon\Theta + \bar{Z}$,

$$\hat{X} = \mathcal{R}_1 - i\varepsilon\mathcal{R}_2 + \varepsilon\mathcal{R}_1 \mathcal{V}_1 + \mathcal{R}_3 e^{i\mathcal{V}_2}$$

acting on functions of $(\mathcal{R}_1, \mathcal{R}_2, \mathcal{R}_3)$. We build a vector space by acting on $\psi_0 = 1$ with the operators $\hat{Z}$, $\hat{\Theta}$, and $\hat{\bar{Z}}$ yielding a vector space of polynomials in $(\mathcal{R}_1, \mathcal{R}_2, \mathcal{R}_3)$.

For an irreducible representation, we want the Casimir operator $Z\bar{Z}$ to act as a scalar, λ say. This gives

$$\mathcal{R}_1 \mathcal{R}_3 e^{i\mathcal{V}_2}\psi = \lambda\psi$$

on the vector ψ. That is,

$$\mathcal{R}_1 \mathcal{R}_3 \psi(\mathcal{R}_1, \mathcal{R}_2 + i, \mathcal{R}_3) = \lambda\psi(\mathcal{R}_1, \mathcal{R}_2, \mathcal{R}_3)$$

which means that the variable $\mathcal{R}_2$ comes in the form of a factor of $e^{\kappa\mathcal{R}_2}$, extending the space to include exponentials, with $\lambda = \mu e^{i\kappa}$, and the condition $\mathcal{R}_1 \mathcal{R}_3 = \mu$, a constant. The action of Θ on an element of the form polynomial in $(\mathcal{R}_1, \mathcal{R}_3)$ times $e^{\kappa\mathcal{R}_2}$ shows that in fact $\kappa = 0$. Thus, the representation space reduces to polynomials in $\mathcal{R}_1$.

We thus consider representations acting on a space with basis ψ_n, $n \in \mathbf{Z}$, with the action of the Lie algebra

$$Z\psi_n = \psi_{n+1}, \quad \bar{Z}\psi_n = \lambda\psi_{n-1}, \quad \Theta\psi_n = in\psi_n$$

We have the recurrence relations given by the action of X on the basis

$$X\psi_n = \psi_{n+1} + \varepsilon n\psi_n + \lambda\psi_{n-1}$$

Observe that one can take $\lambda > 0$ to equal 1 by the scaling $\lambda^{n/2}f(x/\sqrt{\lambda})$. So we consider the representation given by the action on the basis:

$$Z\psi_n = \psi_{n+1}, \quad \bar{Z}\psi_n = \psi_{n-1}, \quad \Theta\psi_n = in\psi_n$$

with the action of X

$$X\psi_n = \psi_{n+1} + \varepsilon n\psi_n + \psi_{n-1}$$

5.5 BESSEL FUNCTIONS

Here we recall some facts about Bessel functions. Watson [52] and Abramowitz & Stegun [1] serve as general references.

The series expansions

$$J_n(z) = \left(\frac{z}{2}\right)^n \sum_{k \geq 0} \frac{(-z^2/4)^k}{k!\,(n+k)!}$$

$$I_n(z) = \left(\frac{z}{2}\right)^n \sum_{k \geq 0} \frac{(z^2/4)^k}{k!\,(n+k)!}$$

with the relations

$$J_{-n}(z) = (-1)^n J_n(z), \qquad I_{-n}(z) = I_n(z)$$

And the generating functions

$$
\begin{aligned}
e^{iz \sin \theta} &= \sum_{n=-\infty}^{\infty} e^{in\theta} J_n(z) \\
e^{z \cos \theta} &= \sum_{n=-\infty}^{\infty} e^{in\theta} I_n(z)
\end{aligned}
\tag{5.5.1}
$$

The generating function for I_n corresponds to the integral formula

$$I_n(z) = \frac{1}{\pi} \int_0^{\pi} e^{z \cos \theta} \, \cos n\theta \, d\theta \tag{5.5.2}$$

And

5.5.1 Proposition. *The expansion*

$$e^{a+b} = \sum_{n=-\infty}^{\infty} \left(\frac{a}{b}\right)^{n/2} I_n(2\sqrt{ab})$$

Proof: Write the generating function for I_n in the form

$$e^{(z/2)(w+w^{-1})} = \sum_{n=-\infty}^{\infty} w^n I_n(z)$$

replacing $e^{i\theta}$ with w. Then, replacing $w = \sqrt{a/b}$ and $z = 2\sqrt{ab}$, the result follows. ∎

Another important feature

5.5.2 Lemma. *For $t \in \mathbf{R}$,*

$$\sum_{n=-\infty}^{\infty} J_n(t)^2 = 1$$

Proof: In eqs. (5.5.1), in the generating function for $J_n(z)$, put $t = z$ real and apply Plancherel's theorem. ∎

5.6 MATRIX ELEMENTS OF e^{sX} AND ORTHOGONALITY

For orthogonality, we take the inner product with Z and $\bar{Z}$ adjoints. Thus, for $p \geq 0$

$$\langle \psi_n, \psi_{n+p} \rangle = \langle \psi_p, \psi_0 \rangle$$

and we have an orthonormal system as long as every ψ_n, $n \neq 0$, is orthogonal to ψ_0.

Remark. Denote:

$$\xi = e^{\varepsilon s}$$

5.6.1 Proposition. *We have the expansion*

$$e^{sX}\,\psi_0 = \sum_{n=-\infty}^{\infty} e^{\varepsilon s n/2}\, I_n \left(2\,\frac{\sinh(\varepsilon s/2)}{\varepsilon/2} \right)\, \psi_n$$

Proof: From the splitting formula, as in eq. (5.3.1),

$$e^{sX}\,\psi_0 = e^{(\xi-1)Z/\varepsilon}\, e^{(1-1/\xi)\bar{Z}/\varepsilon}\, \xi^{-i\Theta}\,\psi_0$$

As $\Theta\psi_0 = 0$, use Proposition 5.5.1, with $a = (\xi-1)Z/\varepsilon$ and $b = (1\ 1/\xi)\bar{Z}/\varepsilon$. The action of $Z/\bar{Z}$ is according to

$$\left(\frac{Z}{\bar{Z}} \right)^{n/2} = \begin{cases} Z^n, & n \geq 0 \\ \bar{Z}^{-n}, & n < 0 \end{cases}$$

And for $2\sqrt{ab}$ observe that

$$\frac{2}{\varepsilon}\sqrt{\xi + \xi^{-1} - 2} = 2\,\frac{\sinh(\varepsilon s/2)}{\varepsilon/2}$$

and the result follows. ∎

Thus

5.6.2 Theorem. *The matrix elements of e^{sX}*

$$\langle e^{sX}\,\psi_n, \psi_m \rangle = e^{\varepsilon s(m+n)/2}\, I_{m-n} \left(2\,\frac{\sinh(\varepsilon s/2)}{\varepsilon/2} \right)$$

Proof: Use the relation $Zf(X) = f(X - \varepsilon)Z$, Proposition 5.3.4, to get

$$e^{sX}\psi_n = e^{sX} Z^n\psi_0 = Z^n e^{sn\varepsilon} e^{sX}\psi_0$$

and bring Z^n across the inner product as $\bar{Z}^n$. Now apply the above Proposition. ∎

Remark. The representation property $M_{ij}(s+t) = \sum M_{il}(s)M_{lj}(t)$ is a version of Graf's addition formula (see [1] or [52]).

Observe that for $s = 0$, $\langle\psi_n, \psi_0\rangle = I_n(0) = 0$ for $n \neq 0$ as required for orthogonality.

5.7 BASIS AND LOMMEL POLYNOMIALS

Given the initial basis vectors ψ_0 and ψ_{-1} the basis ψ_n will be given as functions of X (written as just the variable x). Writing

$$\psi_n(x) = A_n(x)\psi_0 + B_n(x)\psi_{-1} \tag{5.7.1}$$

thus A_n and B_n are the fundamental solutions to the recurrence for ψ_n, i.e., they satisfy the recurrence with the initial conditions

$$A_0 = 1 \qquad B_0 = 0$$
$$A_{-1} = 0 \qquad B_{-1} = 1$$

We recall that the *Lommel polynomials*, here denoted by $R_n(\xi, \nu)$, satisfy the recurrence

$$2(n + \nu)\xi^{-1}R_n = R_{n+1} + R_{n-1}$$

with the initial conditions $R_{-1} = 0$, $R_0 = 1$. The R_n may be given explicitly in the form

$$R_n(\xi, \nu) = \sum_{k=0}^{[n/2]} \binom{n-k}{k}(-1)^k \frac{\Gamma(\nu + n - k)}{\Gamma(\nu + k)} (2/\xi)^{n-2k}$$

Comparing with the recurrence for the ψ_n,

$$A_n(x) = R_n(-2/\varepsilon, -x/\varepsilon)$$

Applying Z to equation (5.7.1) gives

$$Z\psi_n = \psi_{n+1} = A_n(x - \varepsilon)Z\psi_0 + B_n(x - \varepsilon)Z\psi_{-1}$$
$$= A_n(x - \varepsilon)\psi_1 + B_n(x - \varepsilon)\psi_0$$

The recurrence says that $\psi_1 = x\psi_0 - \psi_{-1}$. Thus,

$$\psi_{n+1} = A_n(x - \varepsilon)(x\psi_0 - \psi_{-1}) + B_n(x - \varepsilon)\psi_0$$

And the coefficients A_n and B_n satisfy

$$B_n(x) = -A_{n-1}(x - \varepsilon)$$
$$A_{n+1}(x) = xA_n(x - \varepsilon) - A_{n-1}(x - 2\varepsilon)$$

5.7.1 Theorem.   The basis is given in terms of Lommel polynomials,
$A_n(x) = R_n(-2/\varepsilon, -x/\varepsilon)$, as

$$\psi_n(x) = A_n(x)\psi_0 - A_{n-1}(x - \varepsilon)\psi_{-1}$$

Remark. In the limit $\varepsilon \to 0$, the recurrence reduces to that for Chebyshev polynomials
in the variable $x/2$ with the corresponding relation

$$\lim_{\varepsilon \to 0} R_n(-2/\varepsilon, -x/\varepsilon) = U_n(x/2)$$

5.8 BASIS AND BESSEL FUNCTIONS

Now, we want the basis to be of the form

$$\psi_n(x) = f_n(x)\psi_0$$

for some functions $f_n(x)$. Then, with $\psi_0 = 1$, we have a realization of the basis as functions
of x.

Recall that $\psi_1 = x\psi_0 - \psi_{-1}$. For $\psi_{-1} = f(x)\psi_0$, the action of Z gives

$$\psi_0 = Z\psi_{-1} = Zf(x)\psi_0 = f(x - \varepsilon)\psi_1$$
$$= f(x - \varepsilon)x\psi_0 - f(x - \varepsilon)f(x)\psi_0$$

Thus

$$xf(x - \varepsilon) - f(x)f(x - \varepsilon) = 1 \tag{5.8.1}$$

Or,

$$f(x) = x - \frac{1}{f(x - \varepsilon)}$$

Recursively, generate the continued fraction

$$f(x) = x - 1\Big/ x - \varepsilon - 1\Big/ x - 2\varepsilon - 1\Big/ x - 3\varepsilon - 1\Big/ \ldots$$

Comparing with the continued fraction ([52], p. 153, p. 303)

$$\frac{J_{\nu-1}(z)}{J_\nu(z)} = \frac{2\nu}{z} - 1\Big/ \frac{2(\nu + 1)}{z} - 1\Big/ \frac{2(\nu + 2)}{z} - 1\Big/ \ldots$$

yields (as before) $z = -2/\varepsilon$, $\nu = -x/\varepsilon$:

$$f(x) = \frac{J_{-x/\varepsilon-1}(-2/\varepsilon)}{J_{-x/\varepsilon}(-2/\varepsilon)}$$

Equation (5.8.1) is seen to be the *Bessel recursion*

$$\frac{2\nu}{z} J_\nu(z) = J_{\nu+1}(z) + J_{\nu-1}(z)$$

5.8.1 Theorem. *The basis vectors are given by*

$$\psi_n(x) = \frac{J_{n-x/\varepsilon}(-2/\varepsilon)}{J_{-x/\varepsilon}(-2/\varepsilon)}$$

Proof: With $A_n(x) = R_n(-2/\varepsilon, -x/\varepsilon)$,

$$\psi_n(x) = A_n(x) - A_{n-1}(x - \varepsilon) \frac{J_{-x/\varepsilon-1}(-2/\varepsilon)}{J_{-x/\varepsilon}(-2/\varepsilon)}$$

The basic feature of Lommel polynomials is the relation

$$J_{\nu+n}(z) = R_n(z, \nu)J_\nu(z) - R_{n-1}(z, \nu + 1)J_{\nu-1}(z)$$

and the result follows. ∎

Remark. To verify the expansion of e^{sz} directly, one can use ([52], p. 142)

$$J_\nu(\lambda z) = \lambda^\nu \sum_{n \geq 0} \frac{(-1)^n(\lambda^2 - 1)^n(z/2)^n}{n!} J_{\nu+n}(z)$$

with $\nu = -x/\varepsilon$ in the expansion of eq. (5.3.1), using this successively for the exponential of Z, then $\bar{Z}$.

Note that the action $Z\psi_n = \psi_{n+1}$ becomes

$$Z\psi_n(x) = \psi_n(x - \varepsilon)Z\psi_0 = \psi_n(x - \varepsilon)\psi_1(x)$$

i.e.,

$$\psi_{n+1}(x) = \psi_n(x - \varepsilon)\psi_1(x)$$

which is readily checked.

5.8.1 Orthogonality

We can verify orthogonality directly. The convolution formula ([52], p. 30)

$$J_n(y+z) = \sum_{k=-\infty}^{\infty} J_k(y)J_{n-k}(z)$$

with the relations $J_n(-z) = (-1)^n J_n(z) = J_{-n}(z)$ gives

$$\delta_{0n} = J_{-n}(0) = \sum_{k=-\infty}^{\infty} J_k(y)J_{-n-k}(-y)$$

$$= \sum_{k=-\infty}^{\infty} J_k(y)J_{n+k}(y)$$

We have, writing $x = -k\varepsilon$,

$$\langle \psi_n \psi_m \rangle = \sum_{k=-\infty}^{\infty} \frac{J_{n+k}(-2/\varepsilon)J_{m+k}(-2/\varepsilon)}{J_k(-2/\varepsilon)^2} J_k(-2/\varepsilon)^2$$

$$= \sum_{k=-\infty}^{\infty} J_{n+k}(-2/\varepsilon)J_{m+k}(-2/\varepsilon)$$

$$= \sum_{k=-\infty}^{\infty} J_{n-m+k}(-2/\varepsilon)J_k(-2/\varepsilon) = \delta_{nm}$$

5.9 LOMMEL RANDOM VARIABLES

Now consider X as a random variable, i.e., the spectral form of the self-adjoint operator X, with corresponding spectral measure in the state ψ_0, i.c., expectation $\langle \cdot \rangle$ means $\langle \psi_0, \cdot \psi_0 \rangle$. Thus,

5.9.1 Proposition. *We have the moment generating function*

$$\langle e^{zX} \rangle = I_0 \left(2\frac{\sinh(\varepsilon z/2)}{\varepsilon/2} \right)$$

Proof: This follows from Theorem 5.6.2 with $m = n = 0$. ∎

And, cf. the calculation above, §5.8.1,

5.9.2 Theorem. *The distribution of X is given by*

$$P(X = \pm m\varepsilon) = J_m(2/\varepsilon)^2$$

for $m \in \mathbf{Z}$.

Proof: The above Proposition gives, writing $\xi = e^{\varepsilon z}$,

$$\sum_{n\geq 0} \frac{4^n \sinh^{2n}(\varepsilon z/2)}{\varepsilon^{2n} n!\, n!} = \sum_{n=0}^{\infty}\sum_{\mu} \binom{2n}{\mu} \xi^{(2n-\mu)/2}\xi^{-\mu/2}(-1)^{\mu}/(\varepsilon^{2n} n!\, n!)$$

$$= \sum_{n=0}^{\infty}\sum_{\mu} \binom{2n}{\mu} \xi^{n-\mu}(-1)^{\mu}/(\varepsilon^{2n} n!\, n!)$$

Since this is symmetric in $z \leftrightarrow -z$, equivalently $\xi \leftrightarrow \xi^{-1}$, we consider $n > \mu$, $n = \mu + k$. This gives for the coefficient of ξ^k:

$$\sum_{\mu} \binom{2\mu + 2k}{\mu}(-1)^{\mu}/(\varepsilon^{2\mu+2k}(\mu + k)!\,(\mu + k)!)$$

According to [52], p. 147,

$$J_{\nu}(t)^2 = \sum_{m\geq 0} \frac{(-1)^m (t/2)^{2\nu+2m}(2\nu + 2m)!}{m!\,(2\nu + m)!\,(\nu + m)!\,(\nu + m)!}$$

and the result follows. ∎

5.9.3 Definition. For $t \in \mathbf{R}$, the random variable X_t has a *Lommel distribution with parameter* t if $P(X_t = \pm m) = J_m(t)^2$, $m \in \mathbf{Z}$.

5.9.4 Proposition. The Lommel random variable X_t has characteristic function

$$\langle e^{izX_t} \rangle = J_0(2t\sin(z/2))$$

Proof: In Theorem 5.9.2, replace $\varepsilon \to 2/t$ and note that then $X = 2X_t/t$. Proposition 5.9.1 reads

$$\langle e^{zX} \rangle = I_0(2t\sinh(z/t))$$

Replacing z by $izt/2$ yields the result. ∎

5.9.5 Corollary. A Lommel variable X_t has mean zero and variance $t^2/2$.

We have included a plot of the Lommel distribution for $t = 70$. The jagged nature of the distribution is particularly interesting.

References

The main reference for Bessel functions is Watson [52]. For special functions in general: Abramowitz & Stegun [1]. See [14] for an initial study of stochastic processes on quantum groups. [21] is a study of the zeros of Lommel polynomials and limiting relations with Bessel functions. For a probabilistic approach to Krawtchouk polynomials, see [23]. The q-Heisenberg algebra is considered in [24]. A Krawtchouk transform is discussed in [20].

Chapter 6   Nilpotent and solvable algebras

For this chapter, we have MAPLE output (collected at the end of the book) for typical cases, including higher-dimensional examples. Thus, the discussion will focus on illustrating various other aspects of the theory: matrix elements, Leibniz function, and Lie response.

Remark. For the Lie response, we denote the input stochastic signal by $w(t)$, indicating the Wiener process, and the response process by $W(t)$. The general form of the response is the same for any stochastic signal. Then, expected values calculated in the examples are for the Wiener process.

NOTE: To distinguish expectation with respect to an underlying w-process from the Hilbert space inner product/expectation coming from the Lie structure, we use a subscript w.

I. Heisenberg algebras

The Heisenberg algebras are very similar to the boson algebras, except that the central element is not necessarily the identity. Thus,

1.1 Definition. Given an integer $N > 0$, the *Heisenberg algebra* of dimension $2N + 1$ is spanned by $\{R_i\}$, $\{L_i\}$, for $1 \leq i \leq N$, and ρ satisfying the commutation rules

$$[L_j, R_i] = \delta_{ij}\rho, \qquad [L_i, \rho] = [\rho, R_i] = 0$$

for all $1 \leq i, j \leq N$.

Thus, a boson realization is given by setting $\rho = hI$, for a fixed scalar h, $R_i = \mathcal{R}_i$, $L_i = h\mathcal{V}_i$.

1.1   THREE-DIMENSIONAL HEISENBERG ALGEBRA

For $N = 1$, we have the three-dimensional Heisenberg algebra, with basis ξ_i satisfying $[\xi_3, \xi_1] = \xi_2$, with ξ_2 central. We will look at the matrix elements for the action of the group on $\mathcal{U}(\mathcal{G})$, the Leibniz function and associated Hilbert space, then the Lie response and Appell systems for Brownian motion.

1.1.1 Matrix elements

From the MAPLE output, we see that the right dual representation is

$$\xi^* = \begin{pmatrix} \partial_1 + A_3\partial_2 \\ \partial_2 \\ \partial_3 \end{pmatrix}$$

Using the principal formula, Ch. 2, Theorem 3.1.1, we find

1.1.1.1 Proposition.   *The general matrix elements for the group are:*

$$\left\langle \begin{matrix} m \\ n \end{matrix} \right\rangle = \begin{cases} \dfrac{A^\Delta}{\Delta!} \, {}_2F_1\left(\begin{matrix} -n_1, -\Delta_2 \\ 1 + \Delta_1 \end{matrix} \,\middle|\, \dfrac{A_1 A_3}{A_2} \right), & \Delta_1 \geq 0 \\[3ex] \begin{pmatrix} n_1 \\ -\Delta_1 \end{pmatrix} \dfrac{A_2^{\Delta_1 + \Delta_2} A_3^{\Delta_3 - \Delta_1}}{(\Delta_1 + \Delta_2)! \, \Delta_3!} \, {}_2F_1\left(\begin{matrix} -n_1 - \Delta_1, -\Delta_2 - \Delta_1 \\ 1 - \Delta_1 \end{matrix} \,\middle|\, \dfrac{A_1 A_3}{A_2} \right), & \Delta_1 < 0 \end{cases}$$

where $\Delta = m - n$.

Proof: The principal formula yields

$$\left\langle \begin{matrix} m \\ n \end{matrix} \right\rangle = (\partial_1 + A_3\partial_2)^{n_1} \partial_2^{n_2} \partial_3^{n_3} \frac{A^m}{m!}$$

Expanding yields

$$\begin{pmatrix} n_1 \\ \lambda \end{pmatrix} \partial_1^{n_1 - \lambda} (A_3\partial_2)^\lambda \frac{A_1^{m_1}}{m_1!} \frac{A_2^{\Delta_2}}{\Delta_2!} \frac{A_3^{\Delta_3}}{\Delta_3!} = \begin{pmatrix} n_1 \\ \lambda \end{pmatrix} \frac{A_1^{\Delta_1 + \lambda}}{(\Delta_1 + \lambda)!} \frac{A_2^{\Delta_2 - \lambda}}{(\Delta_2 - \lambda)!} \frac{A_3^{\Delta_3 + \lambda}}{\Delta_3!}$$

which gives the indicated result for $\Delta_1 \geq 0$. For the second case, write $\lambda = -\Delta_1 + \mu$, then rearrange to get the form given.   ∎

The recurrence formulas for these matrix elements yield the *contiguous relations* for the hypergeometric functions. From the **matrec** routine, we have, e.g., from the first row,

$$\left\langle \begin{matrix} m \\ n + e_1 \end{matrix} \right\rangle = m_1 \left\langle \begin{matrix} m - e_1 \\ n \end{matrix} \right\rangle + A_3\, m_2 \left\langle \begin{matrix} m - e_2 \\ n \end{matrix} \right\rangle$$

The action of $\mathcal{R}_2$ is clear from the principal formula as it is an additional factor of ∂_2. The action of $\mathcal{R}_3$ itself only affects the factor in front.

1.1.2 Group law, Leibniz function, and Hilbert space

We can read the group law right off from the grp output. Thus, setting $R = \xi_1$, $\rho = \xi_2$, $L = \xi_3$,

1.1.2.1 Proposition.   *The Leibniz rule is*

$$e^{BL} e^{VR} = e^{VR} e^{BV\rho} e^{BL}$$

and the Leibniz function is $\Upsilon_{BV} = e^{tBV}$ *for the induced representation with* $L\Omega = 0$, $\rho\Omega = t\Omega$.

The Hilbert space in this case is usually called the *Fock space.*   Expanding Υ shows that it has orthogonal basis $R^n \Omega$ with squared norms $\gamma_n = t^n n!$.

1.1.3 Splitting formula and canonical Appell system

From the right or left dual, the splitting lemma, Ch. 2, §II, yields differential equations for the coordinates of the second kind:

$$\dot{A}_1 = \alpha_1, \quad \dot{A}_2 = \alpha_1 A_3 + \alpha_2, \quad \dot{A}_3 = \alpha_3$$

Thus, solving and setting $s = 1$,

$$e^{\alpha_1 \xi_1 + \alpha_2 \xi_2 + \alpha_3 s \xi_3} = g(\alpha_1, \alpha_2 + \alpha_1 \alpha_3 / 2, \alpha_3)$$

Thus,

1.1.3.1 Proposition.   *The element* $X = R + L$ *has a Gaussian distribution with mean zero and variance* t.

Proof: From the splitting formula,

$$e^{zX} = e^{zR} e^{z^2 \rho / 2} e^{zL}$$

With $\rho\Omega = t\Omega$, taking the expected value, the L and R factors drop out.   ∎

1.1.4 Lie response and Wiener process

From the group law, Theorem 2.2 of Ch. 3 yields equations for the Lie response. In the group law (given on the MAPLE output), substitute dw_i for B_i and subtract off A_i to find

$$dW_1 = dw_1 , \quad dW_2 = dw_2 + W_3\, dw_1 , \quad dW_3 = dw_3$$

Since ξ_1, ξ_3 generate the Lie algebra, it is sufficient to take $w_2 = 0$. Thus, the process on the group has the form

$$W_1(t) = w_1(t), \quad W_2(t) = \int_0^t w_3\, dw_1 , \quad W_3(t) = w_3(t)$$

As in Theorem 2.4 of Ch. 3,

$$\langle g(W(t)) \rangle_w = \exp\left(\frac{t}{2}(\xi_1^2 + \xi_3^2) \right)$$

Now, fix $h > 0$, and use the realization $\xi_3 = ihD$, $\xi_1 = X$, multiplication by x, on functions $f(x)$. The commutator yields $\xi_2 = ihI$. And $H(\xi) = \frac{1}{2}(X^2 - h^2 D^2)$, the Hamiltonian for the harmonic oscillator. The eigenfunctions are given by (cf. Volume 1, pp. 30, 117):

$$H\phi_n = h(n + \tfrac{1}{2})\phi_n , \quad \phi_n = H_n\left(x\sqrt{\frac{2}{h}} \right) e^{-x^2/(2h)}$$

where H_n are Hermite polynomials (as in Ch. 1, §2.1, ex. 2). Therefore,

$$\langle e^{W_1 X}\, e^{ihW_2}\, e^{ihW_3 D}\, \phi_n(x) \rangle_w = e^{th^2(n+1/2)^2/2}\, \phi_n(x)$$
$$= \langle e^{W_1 x}\, e^{ihW_2}\, \phi_n(x + ihW_3) \rangle_w$$

yielding correlation functions for the W process.

On the other hand, $L = -D^2/2$, $R = X^2/2$, $\Lambda = -(XD + \tfrac{1}{2})$ satisfy the commutation relations of sl(2) as in Ch. 5, §4.1. Thus, the splitting formula, Ch. 5, 4.1.4, yields, with $H = R + h^2 L$,

$$e^{tH} = e^{(\tanh ht)x^2/(2h)} (\operatorname{sech} ht)^{(XD+1/2)} e^{-(\tanh ht)D^2/2}$$

Applying this to x^n yields

1.1.4.1 Proposition.   *For Brownian motion on the Heisenberg group:*

$$\langle e^{W_1 x}\, e^{ihW_2} (x + ihW_3)^n \rangle_w = e^{(\tanh ht)x^2/(2h)} (\operatorname{sech} ht)^{1/2} H_n(x \operatorname{sech} ht, \tanh ht)$$

Proof: This follows from the relation $e^{-\alpha D^2/2} x^n = H_n(x, \alpha)$ for the Hermite polynomials and the scaling formula $a^{XD} f(x) = f(ax)$. ∎

Remark. Recall here the relation $H_n(x, \alpha) = \alpha^{n/2} H_n(x/\sqrt{\alpha})$.

This illustrates some of the applications of this theory to classical problems.

1.2 GENERAL HEISENBERG ALGEBRA

Now we look at the general Heisenberg algebras. Fix $N > 0$, $d = 2N + 1$. We take the basis ξ_i so that ξ_{N+1} spans the center, with $[\xi_{N+1+i}, \xi_i] = \xi_{N+1}$. The MAPLE output is for $N = 3$, the seven-dimensional algebra. It already shows the features of the general case.

1.2.1 Matrix elements

The right dual is given by $\xi_i^* = \partial_i + A_{N+1+i}\partial_{N+1}$, for $1 \leq i \leq N$, with the remaining $\xi_i^* = \partial_i$. A calculation similar to that for the $N = 1$ case yields, for $\Delta = m - n \geq 0$,

$$\left\langle {m \atop n} \right\rangle = \frac{A^\Delta}{\Delta!} F_A \left({-(n_1, \ldots, n_N), -\Delta_{N+1} \atop (1 + \Delta_1, \ldots, 1 + \Delta_N)} \,\middle|\, \frac{1}{A_{N+1}} (A_1 A_{N+2}, \ldots, A_N A_{2N+1}) \right)$$

where the Lauricella function F_A is defined by the series

$$F_A \left({a, t \atop b} \,\middle|\, \zeta \right) = \sum_{n \geq 0} \frac{(a)_n (t)_{|n|}}{(b)_n n!} \zeta^n$$

for N-component a, b, ζ and single variable t, where, $(a)_n = \prod (a_j)_{n_j}$, and similarly for b.

1.2.2 Group law, Leibniz function, and Hilbert space

With $R_i = \xi_i$, $L_i = \xi_{N+i+1}$, $1 \leq i \leq N$, $\rho = \xi_{N+1}$, we have the group law

$$g(V, H, B)g(V', H', B') = g(V + V', H + H' + B^\dagger V', D + D')$$

and the Leibniz rule

$$e^{BL} e^{VR} = g(V, B^\dagger V, B)$$

Thus, the Leibniz function is $e^{tB^\dagger V}$ and the basis $R^n \Omega$ is an orthogonal basis with squared norms $t^{|n|} n!$.

For the canonical Appell system, we observe that the variable $X = z_\mu X_\mu$, with $X_i = R_i + L_i$, is a sum of commuting independent variables, i.e. $\langle e^X \rangle = \prod \langle e^{z_i X_i} \rangle$, since only central terms are involved.

Similarly, for the Lie response, the group law gives, taking $w_{N+1}(t) = 0$ in the w-process,

$$W_{N+1}(t) = \sum_{i=1}^{N} \int_0^t w_{N+1+i} \, dw_i$$

with $W_j(t) = w_j(t)$ for $j \neq N + 1$. Everything factors into a product of N commuting 3-dimensional subalgebras:

$$g(W(t)) = \prod_{i=1}^{N} \left(e^{W_i \xi_i} \, e^{W_{N+1}^{(i)} \xi_{N+1}} \, e^{W_{N+1+i} \xi_{N+1+i}} \right)$$

where $W_{N+1}^{(i)} = \int_0^t w_{N+1+i} \, dw_i$, with

$$\langle g(W(t)) \rangle_w = \exp \left(\frac{t}{2} \sum_{i=1}^{N} (\xi_i^2 + \xi_{N+1+i}^2) \right)$$

The correlation functions are thus products of the $N = 1$-dimensional functions given according to Proposition 1.1.4.1.

II.   Type-H Lie algebras

A. Kaplan introduced a class of nilpotent Lie groups which arise naturally from the notion of composition of quadratic forms. Since this family of groups is closely related to (and includes) the Heisenberg group, they are referred as *groups of Heisenberg type* or simply of *type-H*. These groups have interesting analytic and geometric properties, e.g., the standard 'Laplacians' on such groups admit fundamental solutions analogous to the case of the Heisenberg group.

Here we recall the basic definitions and properties and look at some of the main features.

Notation. For an array $(a_1, \ldots, a_r)$, $|a| = (\sum a_j^2)^{1/2}$.

2.1 DEFINITIONS

Starting with $\mathcal{Z}, \mathcal{V}$ vector spaces over $\mathbf{R}$, the corresponding two-step nilpotent Lie algebra $\mathcal{G} = \mathcal{V} \oplus \mathcal{Z}$ (the symbol $\oplus$ here denoting orthogonal direct sum) is determined by a mapping $j: \mathcal{Z} \to \mathrm{End}\, \mathcal{V}$ with the properties, for $Y \in \mathcal{Z}$:

$$|j(Y) X| = |Y| |X|$$
$$j(Y)^2 = -|Y|^2 I$$
$$\langle Y, [X, X'] \rangle = \langle j(Y) X, X' \rangle$$

where $\langle \, , \rangle$ denotes a given inner product on $\mathcal{G}$. The space $\mathcal{Z}$ is the center of the algebra. Note that $j|_{\mathcal{V}} = 0$.

Now fix orthonormal bases $\{X_k\}$, $\{Z_k\}$ of $\mathcal{V}$ and $\mathcal{Z}$ respectively. Let

$$J_k = j(z_k)$$

Denote

$$\dim \mathcal{V} = r$$
$$\dim \mathcal{Z} = r'$$

Define the matrix elements

$$J_k^{ij} = \langle J_k X_j, X_i \rangle$$

For each non-zero $z \in \mathcal{Z}$, $j(z)$ is surjective. For $z = c_\lambda z_\lambda \in \mathbf{z}$, $j(c_\lambda z_\lambda) = c_\lambda J_\lambda$. Each J_k is an isometry and $J_k^2 = -I$, i.e., J_k is skew-symmetric. The J's satisfy the Clifford algebra relations

$$J_k J_l + J_l J_k = -2\delta_{kl} I \tag{2.1.1}$$

The following useful proposition is immediate.

2.1.1 Proposition. *For any scalars* $c = \{c_j\}$,

$$c_\lambda J_\lambda^{ip} c_\mu J_\mu^{jp} = |c|^2 \delta_{ij}$$
$$c_\lambda J_\lambda^{ip} c_\mu J_\mu^{pj} = -|c|^2 \delta_{ij}$$

We now find the basic commutation relations. First, calculate

$$\langle z_r, [X_k, X_l] \rangle = \langle J_r X_k, X_l \rangle = J_r^{lk}$$

That is, $[X_k, X_l] = z_\mu J_\mu^{lk}$. Denoting this by $z \cdot J^{lk}$, thus,

2.1.2 Proposition. *The adjoint representation is given by*

$$(\check{X}_k)_{ij} = J_i^{jk}, \quad \check{z}_k = 0$$

where the action of $\check{X}_k$ *is restricted to* $\mathcal{V}$ *and the range is restricted to* $\mathcal{Z}$. *The Kirillov form on* $\mathcal{V} \times \mathcal{V}$, *as a function of* z, *is*

$$K(z) = -z \cdot J$$

In block form, the Kirillov form is

$$K(z) = \begin{matrix} & X & z \\ X \\ z \end{matrix} \begin{pmatrix} z \cdot J^\dagger & 0 \\ 0 & 0 \end{pmatrix}$$

where the decomposition is explicitly indicated.

Remark. Recall the ordering symbol θ_{ij} equal to 1 for $i < j$, 0 otherwise.

2.2 DUAL REPRESENTATIONS

In accordance with the decomposition $\mathcal{G} = \mathcal{V} \oplus \mathcal{Z}$, we denote the coordinates of the first kind by (α, β) and those of the second kind by (A, B): $X = \sum \alpha_i X_i + \sum \beta_k z_k$,

$$g(A, B) = e^{A_1 X_1} \cdots e^{A_r X_r} e^{B_1 z_1} \cdots e^{B_{r'} z_{r'}}$$

We will find the dual representations directly.

2.2.1 Proposition.   *For the left action,*

$$X_i^{\dagger} = \partial/\partial A_i + A_\mu J_\lambda^{\mu i} \theta_{\mu i} \partial/\partial B_\lambda$$
$$z_k^{\dagger} = \partial/\partial B_k$$

For the action on the right

$$X_i^* = \partial/\partial A_i + A_\mu J_\lambda^{i \mu} \theta_{i \mu} \partial/\partial B_\lambda$$
$$z_k^* = \partial/\partial B_k$$

The double dual has the form

$$\hat{X}_i = \mathcal{R}_i + \mathcal{R}_\lambda J_\lambda^{\mu i} \theta_{\mu i} \mathcal{V}_\mu \,, \quad \hat{z}_k = \mathcal{R}_k$$

Proof: A sample computation for each case suffices to illustrate the structure,

$$X_3 e^{A_2 X_2} = e^{A_2 X_2} e^{-A_2 \mathrm{ad} \, X_2} X_3 = e^{A_2 X_2} (X_3 + A_2 z \cdot J^{23})$$
$$e^{A_3 X_3} X_2 = (X_2 + A_3 z \cdot J^{23}) e^{A_3 X_3}$$

■

We can write the pi-matrices in block form as follows:

$$\pi^{\dagger} = \begin{pmatrix} I & A_\mu J^{\mu''} \theta_{\mu''} \\ 0 & I \end{pmatrix} \,, \quad \pi^* = \begin{pmatrix} I & A_\mu J''^{\mu} \theta_{''\mu} \\ 0 & I \end{pmatrix} \tag{2.2.1}$$

where the dots refer to column indices and the "s to row indices.

Next, since the z_k are central, we take $X = \alpha_\mu X_\mu$ for the splitting formula.

2.2.2 Theorem.   *The splitting formula is*

$$e^{\alpha_\mu X_\mu} = g(\alpha, \tfrac{1}{2} \alpha^{\dagger} J_. \theta \alpha)$$

where $(J_k \theta)_{ij} = J_k^{ij} \theta_{ij}$ is the upper-triangular part of J_k.

Proof: We consider e^{sX} as in the splitting lemma. Since the Lie algebra is 2-step nilpotent, the splitting lemma yields the coordinates $A(s), B(s)$ depending at most on s to order two, cf. the Heisenberg case, as can be seen from the form of the pi-matrices, eq. (2.2.1). From the pi-matrices, it is clear that $A_i = \alpha_i$, $1 \leq i \leq r$. From Theorem 1.1 of Ch. 2,

$$B_i(s) = s\beta_i - \frac{s^2}{2}\alpha_\lambda\alpha_\mu\theta_{\lambda\mu}c^i_{\lambda\mu}$$

As remarked in Proposition 2.1.2, the J's and c's differ by a sign. ∎

We now compute the group law.

2.2.3 Proposition. *The group law is*

$$g(A,B)g(A',B') = g(A + A', B + B' + A'_\lambda A_\mu J^{\lambda\mu}_.\theta_{\lambda\mu})$$

Proof: We apply Lemma 2.3.1 of Ch. 2. By the right dual representation, moving the z factors through,

$$g(A,B)g(A',B') = g(A', X^*)g(A, B + B')$$

And the result follows from the form of the right dual as given in Proposition 2.2.1. ∎

2.3 DISTRIBUTION OF X

For type-H algebras we need to have a decomposition into R's and L's and an inner product with respect to which they are adjoints. This can be done by an appropriate change-of-basis. Namely, take the cyclic vector Ω such that $z_j\Omega = \tau_j\Omega$ for some given scalars τ_j. Then the Lie bracket on $\mathcal{G}$ determines a skew-symmetric form B on $\mathcal{G} \times \mathcal{G}$ via $B(X,Y) = [X,Y]\Omega$. Observe that B is non-degenerate. With $B(X_j, X_i) = \tau_\mu J^{ij}_\mu$, for any vector ξ, by eq. (2.1.1),

$$\langle \tau_\lambda J_\lambda \xi, \tau_\mu J_\mu \xi \rangle = \tau_\mu\tau_\mu\langle J_\mu\xi, J_\mu\zeta\rangle$$
$$= |\tau|^2|\xi|^2$$

By Darboux' lemma, make an orthogonal change of variables such that B takes the form of a direct sum of 2×2 matrices of the form $\begin{pmatrix} 0 & h_i \\ -h_i & 0 \end{pmatrix}$. Arrange this so that the h's are all positive. Associated to each block, define pairs of elements L_i, R_i such that $B(L_i, R_i) = h_i$. This yields a decomposition into 'baby Heisenberg' algebras. The representation space has basis $R^n\Omega$. Now we have our Fock-type space. As we saw for the Heisenberg case, Proposition 1.1.3.1,

2.3.1 Proposition. *The elements $R_i + L_i$ form a family of independent Gaussian random variables.*

In general, from the splitting formula:

$$\langle e^{\zeta\alpha_\mu X_\mu} \rangle = e^{\frac{1}{2}\zeta^2\tau_\mu J^{\rho\sigma}_\mu \alpha_\rho\alpha_\sigma \theta(\rho,\sigma)}$$

the quantity in the exponent required to be positive.

2.4 LIE RESPONSE

Here we take the input process $w(t)$ with the components $w_i(t) = 0$, for $i > r$. Then from the group law, Proposition 2.2.3, we have the equations

$$dW_i = dw_i, \qquad 1 \le i \le r$$

$$dW_i = w_\mu J_i^{\lambda\mu} dw_\lambda \theta_{\lambda\mu}, \qquad i > r$$

Therefore the process on the group is of the form

$$g\left(w(t), \int_0^t w_\mu J_.^{\lambda\mu} dw_\lambda \theta_{\lambda\mu}\right)$$

As for the Heisenberg case, using a specific realization of the algebra, we can compute Appell systems corresponding to the process by applying this to functions such as x^n. One realization is given by replacing $\mathcal{R}$ and $\mathcal{V}$ in the double dual by x and D:

$$X_i = x_i + \tau_\lambda J_\lambda^{\mu i} \theta_{\mu i} D_\mu \tag{2.4.1}$$

where we use x as multiplication by the variable x to avoid confusion with the $X's$. The z's are mapped to scalars τ_i.

2.5 HAMILTON'S EQUATIONS

As in §3.8 of the Introduction and in Ch. 2, §2.8, we would like to find the equations for a flow of the form $e^{tH} \xi e^{-tH}$ where H is a function on the algebra $\mathcal{G}$, e.g., a polynomial in the X variables (or a formal power series) or a combination of exponentials. We can use the observations of Ch. 2, §2.8, with the dual representations given above. As well, let us see how this works directly. Thinking of an element of the enveloping algebra, we can consider, a bit more generally, a Hamiltonian of the form $H(X_1, \ldots, X_r) = H_1(X_1) \cdots H_r(X_r)$. Then, as in the calculation of the dual representations,

$$[H, X_i] = (-\operatorname{ad} X_i)(H)$$

$$= -\sum H_1(X_1) \cdots [X_i, H(X_j)] \cdots H(X_m)$$

$$= \sum_j [X_j, X_i] \frac{\partial H}{\partial X_j}$$

Thus,

2.5.1 Theorem. *Hamilton's equations are given by:*

$$\dot{X}_i = [X_\mu, X_i] \frac{\partial H}{\partial X_\mu} \qquad \text{or} \qquad \dot{X}_i = z \cdot J^{i\mu} \frac{\partial H}{\partial X_\mu}$$

$$\dot{z}_i = 0 \qquad\qquad\qquad \dot{z}_i = 0$$

These can be written compactly in the form:

$$\dot{\mathbf{X}} = z \cdot J \nabla H$$

$$\dot{z} = 0$$

The skew-symmetric form $z \cdot J$ gives a Poisson structure—and thus an associated "classical mechanics."

2.5.1 Heat polynomials

Now consider the 'Laplacian' for this case: $H = \frac{1}{2} \sum X_j^2$. From Theorem 2.5.1, we have, via Proposition 2.1.1, that $X(t) = e^{tH} X e^{-tH}$ satisfies $\ddot{X} = -|z|^2 X$. Thus,

$$X(t) = e^{tH} X e^{-tH} = (\cos t|z|) X + \frac{\sin t|z|}{|z|} z \cdot J X$$

which we write in the form $X(t) = \kappa X + \sigma z \cdot JX$. (Note that it is convenient here to think of the variables z_k realized as scalars.) Now we observe

2.5.1.1 Proposition. $\xi_i = \kappa X_i$, $\delta_i = \sigma z \cdot J^{i\mu} X_\mu$ generate a Heisenberg algebra. In fact,

$$[\delta_i, \xi_j] = \sigma \kappa |z|^2 \delta_{ij}$$

Proof: From Proposition 2.1.1, $z \cdot J^{i\mu} z \cdot J^{j\mu} = |z|^2 \delta_{ij}$, so that

$$[\sigma z \cdot J^{i\mu} X_\mu, \kappa X_j] = \sigma \kappa |z|^2 \delta_{ij}$$

as required. ∎

Thus, $X_i(t)$ is of the form $\xi_i + \delta_i$, with δ_i, ξ_i, generating a Heisenberg algebra. Recall the Hermite moment polynomials, $h_k(x,t)$, with generating function

$$e^{ax + a^2 t/2} = \sum_{k \geq 0} \frac{a^k h_k(x,t)}{k!}$$

We find

2.5.1.2 Proposition. With $\kappa = \cos t|z|$ and $\sigma = |z|^{-1} \sin t|z|$,

$$X_i(t)^k = \sum_j \binom{k}{j} \kappa^{k-j} X_i^{k-j} h_j(\sigma z \cdot J^{i\mu} X_\mu, \sigma \kappa |z|^2)$$

Proof: By the splitting formula in the Heisenberg case,

$$e^{a X_i(t)} = e^{a \xi_i} e^{a \delta_i} e^{\frac{a^2}{2} \sigma \kappa |z|^2}$$

Thus,

$$X_i(t)^k = \sum_j \binom{k}{j} \xi_i^{k-j} h_j(\delta_i, \sigma \kappa |z|^2)$$

and the result follows. ∎

To get functions of x, say, using the realization of eq. (2.4.1), substitute for the X_i and apply to $\Omega = 1$. These yield explicit solutions to the heat equation for type-H groups.

2.5.2 Heat flow and sl(2)

For another realization in terms of (x, D) variables, set

$$X = x - \tfrac{1}{2}z \cdot JD = x + \mathcal{A}D$$

i.e., $\mathcal{A} = -\tfrac{1}{2}z \cdot J$. The requisite commutation relations are readily verified.

So consider $H = \tfrac{1}{2}\sum X_j^2$, with $X = x + \mathcal{A}D$. Introduce the variables

$$\delta_i = \mathcal{A}^{i\mu}D_\mu$$
$$R_i = \tfrac{1}{2}x_i^2$$
$$L_i = \tfrac{1}{2}\delta_i^2$$

Thus, $[\delta_i, x_j] = \mathcal{A}^{ij}$. And check

2.5.2.1 Proposition. *The commutation relations are:*

$$[L_k, R_j] = (x_j\delta_k + \tfrac{1}{2}\mathcal{A}^{kj})\mathcal{A}^{kj}$$
$$[L_i, x_j\delta_k] = \mathcal{A}^{ij}\delta_i\delta_k$$
$$[x_i\delta_j, R_k] = \mathcal{A}^{jk}x_ix_k$$

Now $X_i = x_i + \delta_i$, with $[\delta_i, x_i] = 0$. Thus H takes the form

$$H = \tfrac{1}{2}\sum(x_i^2 + \delta_i^2) + x_\mu\delta_\mu = \sum(R_j + L_j) + x_\mu\delta_\mu$$

where, as follows by the skew-symmetry of $\mathcal{A}$, $[\sum(R_j + L_j), x_\mu\delta_\mu] = 0$. From the above Proposition we have that

$$\Delta = \frac{2}{|z|}\sum L_j = \frac{1}{|z|}\sum(\mathcal{A}^{j\mu}D_\mu)^2$$
$$R = \frac{2}{|z|}\sum R_j = \frac{1}{|z|}\sum x_j^2$$
$$\rho = \frac{4}{|z|^2}[\sum L_j, \sum R_i] = \frac{4}{|z|^2}(x_\lambda\delta_\mu + \tfrac{1}{2}\mathcal{A}^{\mu\lambda})\mathcal{A}^{\mu\lambda}$$

satisfy the sl(2) commutation relations

$$[\Delta, R] = \rho, \quad [\rho, R] = 2R, \quad [\Delta, \rho] = 2\Delta$$

(where in Ch. 5, we use $L \leftrightarrow -\Delta$, $\Lambda \leftrightarrow -\rho$, $R \leftrightarrow R$.) Thus, as in the Heisenberg case, the splitting formula for sl(2), Ch. 5, Proposition 4.1.4, solves the heat flow for groups of type-H.

2.6 MATRIX ELEMENTS

Here we just comment briefly on the structure of the matrix elements. The basis for the enveloping algebra is taken in the form $[\![n\,n']\!] = X^n z^{n'}$, with the group action

$$g[\![n\,n']\!] = \sum_{m,m'} \left\langle \begin{matrix} m\,m' \\ n\,n' \end{matrix} \right\rangle [\![m\,m']\!]$$

The principal formula says

$$\left\langle \begin{matrix} m\,m' \\ n\,n' \end{matrix} \right\rangle = (X^*)^n (z^*)^{n'} \frac{A^m}{m!} \frac{B^{m'}}{m'!}$$

From Proposition 2.2.1, we see that the matrix elements will be generalized multivariate functions, analogous to the Lauricella functions for the Heisenberg case, §1.2.1, with the calculations carried out in a similar manner.

III. Upper-triangular matrices

Here we consider upper-triangular matrices. Nilpotent algebras can be realized using strictly upper-triangular matrices, while solvable algebras can be expressed using upper-triangular matrices, i.e., the diagonal is involved.

3.1 FINITE-DIFFERENCE ALGEBRA REVISITED

The algebra of 2×2 upper-triangular matrices is the same as the finite-difference algebra expressed in a different basis. Here we have given MAPLE output for the algebra in the 'standard basis.' We content ourselves with computing the matrix elements of the group acting on the universal enveloping algebra.

3.1.1 Matrix elements

The right dual is given by

$$\xi^* = \begin{pmatrix} \partial_1 + A_2 \partial_2 \\ e^{A_3} \partial_2 \\ \partial_3 \end{pmatrix}$$

By the principal formula,

$$\left\langle \begin{matrix} m \\ n \end{matrix} \right\rangle = (\partial_1 + A_2 \partial_2)^{n_1} (e^{A_3} \partial_2)^{n_2} \partial_3^{n_3} \frac{A^m}{m!}$$

$$= \binom{n_1}{\lambda} \partial_1^{n_1 - \lambda} (A_2 \partial_2)^{\lambda} \frac{A_1^{m_1}}{m_1!} \frac{A_2^{\Delta_2}}{\Delta_2!} \frac{e^{n_2 A_3} A_3^{\Delta_3}}{\Delta_3!}$$

which yields, for $\Delta_1 \geq 0$,

$$\left\langle \begin{matrix} m \\ n \end{matrix} \right\rangle = e^{n_2 A_3} \frac{A^{\Delta}}{\Delta!} \, {}_1F_1 \left(\begin{matrix} -n_1 \\ 1 + \Delta_1 \end{matrix} \, \middle| \, -A_1 \Delta_2 \right)$$

3.2 OSCILLATOR ALGEBRA

The oscillator algebra adjoins the number operator to the Heisenberg algebra. For details concerning the associated Hilbert space and canonical Appell structure, the reader is referred to Volume 1, esp. pp. 147–150.

See the end of the book for the MAPLE output.

3.3 ABCD-TYPE ALGEBRAS

The 'ABCD' name suggests a sequential, or flag, structure to the algebra. First take $\xi_d = D$. Then generate the Lie algebra by successive bracketing starting with $\xi_1 = X^{d-2}/(d-2)!$. This yields a $d-1$-step nilpotent algebra. For $d = 4$, this is our original 'ABCD' algebra, for which the MAPLE output is given. Observe from the Kirillov form that ξ_3 generates the center. This algebra is an example of an algebra having a symmetric flag.

3.3.1 Lie response

The procedure grp shows the group law directly, as seen by comparing the form of the group element and the algebra element X. Thus, with ξ_4, ξ_1 generating the Lie algebra,

3.3.1.1 Proposition. *For the ABCD algebra, the Lie response has the form*

$$W_1(t) = w_1(t),\ W_2(t) = \int_0^t w_4\,dw_1,\ W_3(t) = \frac{1}{2}\int_0^t w_4^2\,dw_1,\ W_4(t) = w_4(t)$$

where the input process is taken with $w_2(t) = w_3(t) = 0$.

The Wiener process on the group satisfies

$$\langle g(W(t))\rangle = \exp\left(\frac{t}{2}(\xi_1^2 + \xi_4^2)\right)$$

and with the realization $\xi_4 = D$, $\xi_1 = \varepsilon x^2/2 + x$, cf. ξ^*, we can find explicit associated Appell systems. Notice that the corresponding Hamiltonian is of the form $\frac{1}{2}(D^2 + (x + \varepsilon x^2/2)^2)$, which is a type of anharmonic oscillator.

3.3.2 Matrix elements

The form of the matrix elements for this case is interesting as it gives some idea of the types of generalized hypergeometric functions that arise. From the right dual and the principal formula:

$$\left\langle {m \atop n} \right\rangle = (\partial_1 + A_4\partial_2 + \tfrac{1}{2}A_4^2\partial_3)^{n_1}(\partial_2 + A_4\partial_3)^{n_2}\partial_3^{n_3}\partial_4^{n_4}\,\frac{A^m}{m!}$$

Notice the Heisenberg subalgebra spanned by $\{\xi_2, \xi_3, \xi_4\}$. Thus, expanding, one form of the matrix elements is given by

$$
\left\langle {m \atop n} \right\rangle = \frac{A^\Delta}{\Delta!} \sum_{j,k} \frac{(-n_1)_{j+k}(-\Delta_1)_j(-\Delta_3)_k}{(1+\Delta_1)_{j+k}\, j!\, k!}
$$

$$
\times \left(\frac{A_1 A_4}{A_2} \right)^j \left(\frac{A_1 A_4^2}{2A_3} \right)^k {}_2F_1 \left({-n_2, k - \Delta_3 \atop 1 + \Delta_1 - j} \,\middle|\, \frac{A_2 A_4}{A_3} \right)
$$

3.4 STRICTLY UPPER-TRIANGULAR ALGEBRA

An interesting class of nilpotent algebras are the full strictly upper triangular matrices of a given size. We have included the MAPLE results for n5. This shows the general pattern for nd. An induced representation is indicated, using ξ_1, ξ_2, ξ_3 as variables, raising operators, and taking $\xi_4 = \rho_0$, mapping to t on Ω.

IV. Affine and Euclidean algebras

In addition to the study of e2 given in Chapter 5, we present some examples from the affine and Euclidean groups.

4.1 AFFINE ALGEBRA

The affine group Aff$d$ is the semi-direct product of GL(d) with the translation group acting on $\mathbf{R}^d$. The corresponding Lie algebra is a direct sum of gl(d) with the generators of the translations, an abelian Lie algebra isomorphic to $\mathbf{R}^d$. A typical element of aff$d$ has the block form

$$
X = \begin{pmatrix} a & v \\ 0 & 0 \end{pmatrix}
$$

where a is $d \times d$ and v is $1 \times d$. Denoting the abelian subalgebra $\mathcal{P}$ and gl(d) by $\mathcal{K}$, we have $\mathcal{G} = \mathcal{P} \oplus \mathcal{K}$ with the group factorization $\mathbf{G} = \mathbf{PK}$. This is of affine-type as seen in Chapter 2, §2.1. From Proposition 2.1.2of Ch. 2, we have a general form for the splitting formula. Thus, induced representations are in accordance with those of the subgroup $\mathbf{K}$.

We proceed to look at some details for aff1. The Kirillov form (see the MAPLE results) shows the commutation relations

$$
[\xi_2, \xi_1] = \xi_1
$$

4.1.1 Matrix elements

From the right dual and the principal formula follows that

$$\left\langle{m\atop n}\right\rangle = (e^{A_2}\partial_1)^{n_1}\partial_2^{n_2}\frac{A^m}{m!}$$

which readily leads to

$$\left\langle{m\atop n}\right\rangle = e^{n_1 A_2}\frac{A^\Delta}{\Delta!}$$

It is interesting to see how the results of matrec apply in this case.

4.1.2 Lie response

The group law is almost immediate from the output of grp. We have

$$g(A)g(A') = g(A_1 + e^{A_2}A_1', A_2 + A_2')$$

Thus the Lie response is

$$W_1(t) = \int_0^t e^{w_2}\,dw_1, \qquad W_2(t) = w_2(t)$$

And for the Wiener process,

$$\langle g(W(t))\rangle = \exp\left(\frac{t}{2}(\xi_1^2 + \xi_2^2)\right)$$

Here we take the realization $\xi_1 = -iD$, $\xi_2 = -XD$. Then

$$\langle e^{-iW_1 D}e^{-W_2 XD}f(x)\rangle = \langle f(e^{-W_2}(x - iW_1))\rangle$$
$$= \exp\left(\frac{t}{2}\left((XD)^2 - D^2\right)\right)f(x)$$

Here the eigenfunctions are Tchebychev polynomials:

$$((XD)^2 - D^2)T_n = n^2 T_n$$

(note: n is a single index). Recall the formulas

$$T_n(x) = \sum_k \binom{n-k}{k}\frac{n/2}{n-k}(-1)^k(2x)^{n-2k}$$

$$x^n = 2^{-n}\sum_k \binom{n}{k}T_{n-2k}(x)$$

Therefore, with $f(x) = x^n$,

$$\langle e^{-nW_2} (x - iW_1)^n \rangle = 2^{-n} \sum_k \binom{n}{k} e^{t(n-2k)^2/2} \, T_{n-2k}(x)$$

Expanding out $T_{n-2k}(x)$ in powers of x and matching coefficients yields the correlation functions for the W_1-process with the exponential of w_2. To get the correlations including w_2 itself, take the realization $\xi_1 = -iD$, $\xi_2 = -XD - a$. Now the eigenfunctions are Gegenbauer polynomials (see, e.g., Volume 1, p. 60),

$$((XD + a)^2 - D^2)C_n^a(x) = (n + a)^2 C_n^a(x)$$

with

$$C_n^a(x) = \sum_k \frac{(a)_{n-k}}{(n - 2k)! \, k!} (-1)^k (2x)^{n-2k}$$

$$x^n = \frac{n! \Gamma(a)}{2^n} \sum_k \frac{a + n - 2k}{k! \, (a + n - k)!} \, C_{n-2k}^a(x)$$

And

$$\langle e^{-aW_2} e^{-nW_2} (x - iW_1)^n \rangle = \frac{n! \Gamma(a)}{2^n} \sum_k \frac{a + n - 2k}{k! \, (a + n - k)!} \, C_{n-2k}^a(x) e^{t(n+a-2k)^2/2}$$

Expanding out the Gegenbauer polynomials and matching powers of x, one then expands in powers of a, using Hermite moment polynomials, to find the joint correlation functions for the W-process and the exponential of w_2.

The underlying reason for the appearance of ultraspherical polynomials is connected with separating the (Euclidean) Laplacian into radial and angular parts:

$$r^2 \Delta = (d - 2) r \frac{\partial}{\partial r} + \left(r \frac{\partial}{\partial r} \right)^2 + \sum \left(\frac{\partial}{\partial \theta_{ij}} \right)^2$$

where $r^2 = \sum x_j^2$, Δ is the usual Laplacian, and θ_{ij} is the angle of rotation in the ij-plane. The Laplacian is built from the generators of the translations, while $r\partial/\partial r$ generates dilations, just the diagonal part of the K subgroup of the affine group in higher dimensions. Notice that the angular part is the 'Laplacian' or Hamiltonian we get if we input the Wiener process into $SO(d)$.

4.2 GENERAL EUCLIDEAN

The Euclidean algebras are formed by specializing the general linear group of the affine case to the orthogonal group. I.e., a Euclidean algebra is the Lie algebra of the group of rotations and translations of Euclidean geometry.

We have included MAPLE output for e3 for illustration.

References

More details for type-H groups may be found in [15]. [13] discusses the nilpotent case. In [24], calculations of the matrix elements for several examples, including the three-dimensional Heisenberg algebra, are given. For stochastic processes, see [22] for further details. The technique of calculating the correlations in the case of the affine group was developed in the setting of quantum groups, see [14].

Chapter 7   Hermitian symmetric spaces

In this chapter, the theory is illustrated by looking at examples from the Hermitian symmetric spaces of types I-III. They are considered in their matrix realizations as domains in complex matrix spaces via the Harish-Chandra embedding as in Chapter 4. The fermion algebra is discussed in the context of the spaces of type II, so($2p$). Examples of MAPLE output for su(2,2), sp(4), and so(6) have been included in the MAPLE section at the back of the book.

I. Basic structures

The Hermitian symmetric spaces are homogeneous spaces of semisimple Lie groups which can be realized as domains in several complex variables. They are symmetric and Hermitian, i.e., at each point the group has an automorphism of order 2, for which the given point is an isolated fixed point, and the space has a complex structure making it a complex manifold. E. Cartan discovered these spaces and described their basic properties. The irreducible ones are of six types. Here we will consider only types I-III. The type I spaces are homogeneous spaces of $SU(p,q)$, $p \geq q$, $p+q \geq 2$, realized as $p \times q$ matrices. Type II spaces are homogeneous spaces of $SO(2p)$, the special orthogonal groups, realized as spaces of $p \times p$ anti-symmetric matrices. The spaces of type III are spaces of $p \times p$ symmetric matrices, corresponding to the symplectic groups $Sp(2p)$.

In these cases, the Lie algebra has the block form of Ch. 4, corresponding to the Cartan decomposition $\mathcal{G} = \mathcal{P} \oplus \mathcal{K} \oplus \mathcal{L}$, $X = \begin{pmatrix} \alpha & v \\ -\beta & \delta \end{pmatrix}$. The homogeneous space is embedded in the group as the factor e^{VR}, equivalently, as the ratio $g_{12}g_{22}^{-1}$ of the block form of the group element

$$g = \begin{pmatrix} I & V \\ 0 & I \end{pmatrix} \begin{pmatrix} E & 0 \\ 0 & D \end{pmatrix} \begin{pmatrix} I & 0 \\ -B & I \end{pmatrix} = \begin{pmatrix} E - VDB & VD \\ -DB & D \end{pmatrix}$$

The Lie algebra has a distinguished element ρ_0. The Cartan decomposition is determined by the eigenspaces of ρ_0. In matrix form, it is a generalization of the matrix $\begin{pmatrix} 1 & 0 \\ 0 & -1 \end{pmatrix}$ of sl(2), specifically

$$\rho_0 = \begin{cases} \begin{pmatrix} qI & 0 \\ 0 & -pI \end{pmatrix}, & \text{for type I} \\ \\ \begin{pmatrix} I & 0 \\ 0 & -I \end{pmatrix}, & \text{for types II and III} \end{cases}$$

where for types II and III, the I's are $p \times p$ blocks. The example $sl(N + 1)$ in Ch. 4, illustrates the type I case, corresponding to su($N,1$).

Taking $\rho_0 = \begin{pmatrix} I & 0 \\ 0 & -I \end{pmatrix}$ for purposes of illustration, we have

1.1 Proposition.   With the block forms $\mathcal{P} = \{ \begin{pmatrix} 0 & v \\ 0 & 0 \end{pmatrix} \}, \mathcal{K} = \{ \begin{pmatrix} \alpha & 0 \\ 0 & \delta \end{pmatrix} \},$

$\mathcal{L} = \{ \begin{pmatrix} 0 & 0 \\ -\beta & 0 \end{pmatrix} \}$, the eigenspace decomposition of ρ_0 is:

$$[\rho_0, K] = 0, \forall K \in \mathcal{K}, \quad [\rho_0, R] = 2R, \forall R \in \mathcal{P}, \quad [\rho_0, L] = -2L, \forall L \in \mathcal{L}$$

Proof: This follows by direct calculation using the block matrices.   ∎

Observe that $\mathcal{K}$ is determined as the centralizer of ρ_0. Thus, the coadjoint orbit of **G** acting by conjugation on ρ_0 is isomorphic to the coset space **G/K**. The symmetric domain is then the coset space **G/KL**. This corresponds precisely to the induced representation of $\mathcal{U}(\mathcal{G})$, with the action on the cyclic vector Ω given by

$$L\Omega = 0, \quad \rho_0\Omega = \tau\Omega, \quad \rho\Omega = 0$$

for $L \in \mathcal{L}$, and $\rho \in \mathcal{K}, \rho \neq \rho_0$. The group action on Ω is

$$e^X \Omega = e^{\tau H_0} e^{VR} \Omega$$

The following Proposition gives some insight into the structure.

1.2 Proposition.   Given ρ_0 with eigenspace decomposition $\mathcal{P}, \mathcal{K}, \mathcal{L}$ and eigenvalues $\lambda, 0, -\lambda$, respectively, $\lambda \neq 0$. Then $\mathcal{P}$ and $\mathcal{L}$ must be abelian and satisfy $[\mathcal{L}, \mathcal{P}] \subset \mathcal{K}$.

Proof: The Jacobi identity gives, for $R, R' \in \mathcal{P}$,

$$[\rho_0, [R, R']] = \lambda[R, R'] + \lambda[R, R'] = 2\lambda[R, R']$$
$$= \lambda[R, R']$$

since $[R, R'] \in \mathcal{R}$. Thus, $[R, R'] = 0$. And similarly for $\mathcal{L}$. Now take $R \in \mathcal{R}, L \in \mathcal{L}$:

$$[\rho_0, [L, R]] = -\lambda[L, R] + \lambda[L, R] = 0$$

so that $[L, R] \in \mathcal{K}$.   ∎

In the Lie algebra, all matrices have trace zero, i.e., are embedded in $\mathrm{sl}(p + q)$ or $\mathrm{sl}(2p)$, and in the block form $\begin{pmatrix} \alpha & v \\ -\beta & \delta \end{pmatrix}$, $\mathrm{tr}\,\alpha = \mathrm{tr}\,\delta = 0$, except for ρ_0. Thus, in the group, all factors have determinant 1, and in the factors $\begin{pmatrix} E & 0 \\ 0 & D \end{pmatrix}$, coming from **K**, $\det E = \det D = 1$, except for the term involving ρ_0. In the $p \times p$ case, with ρ_0 as above, we have

$$e^{\rho_0 H_0} = \begin{pmatrix} e^{H_0} & 0 \\ 0 & e^{-H_0} \end{pmatrix}$$

so that $\det D = e^{-pH_0}$, and as in Lemma 3.1.4 of Ch. 4, $H_0 = (\det D)^{-1/p}$. At this point we proceed to look at features of these spaces in some detail. The discussion in Ch. 4, §3.1, for $\mathrm{sl}(N + 1)$ will be a guide for the more general setting here.

II. Space of rectangular matrices

The space corresponding to the group $SU(p, q)$, is the space of $p \times q$ matrices having block form

$$\begin{pmatrix} \alpha & v \\ -\beta^\dagger & \delta \end{pmatrix}$$

with α: $p \times p$, v and β: $p \times q$, and δ: $q \times q$.

This space has a Cartan decomposition according to

$$\rho_0 = \begin{pmatrix} qI & 0 \\ 0 & -pI \end{pmatrix}$$

with

$$[\rho_0, R] = (p + q)R, \quad [\rho_0, L] = -(p + q)L$$

for $R \in \mathcal{P}, L \in \mathcal{L}$, and $\mathcal{K}$ the centralizer of ρ_0. Here we use double-indexed raising and lowering operators, and the Jordan map in boson variables gives:

$$\begin{aligned} R_{ij} &= \mathcal{R}_i \mathcal{V}_{j+p} \\ L_{ij} &= -\mathcal{R}_{j+p} \mathcal{V}_i \end{aligned}, \quad 1 \le i \le p, 1 \le j \le q$$

Calculate,

$$[L_{ij}, R_{kl}] = \delta_{jl} \mathcal{R}_k \mathcal{V}_i - \delta_{ik} \mathcal{R}_{j+p} \mathcal{V}_{l+p}$$

and

$$[L_{ij}, R_{ij}] = \mathcal{R}_i \mathcal{V}_i - \mathcal{R}_{j+p} \mathcal{V}_{j+p}$$

Thus, summing over i and j,

2.1 Proposition.   *The element ρ_0 satisfies*

$$\rho_0 = \sum_{\substack{1 \le i \le p \\ 1 \le j \le q}} [L_{ij}, R_{ij}]$$

2.1 COHERENT STATES

Generally, X has the form

$$X = \begin{pmatrix} \alpha & v \\ -\beta^\dagger & \delta \end{pmatrix}$$

For ρ_0, $\delta = -pI$, while for $X \neq \rho_0$, $\operatorname{tr} \delta = 0$. In the induced representation, elements of $\mathcal{K}$ and $\mathcal{L}$ act as zero on Ω except

$$\rho_0 \Omega = pqt\Omega$$

Then

$$g\Omega = \begin{pmatrix} E - VDB^\dagger & VD \\ -DB^\dagger & D \end{pmatrix} \Omega = e^{tH} e^{VR} \Omega$$

so that, as remarked in §I,

$$e^{\rho_0 H_0} = \begin{pmatrix} e^{qH_0} & 0 \\ 0 & e^{-pH_0} \end{pmatrix}$$

Hence, $\det D = e^{-pqH_0}$, or $H_0 = (\det D)^{-1/(pq)}$. Thus,

$$e^{\rho_0 H_0} \Omega = e^{pqtH_0} \Omega = (\det D)^{-t}$$

From Ch. 4, §1.8, cf. Ch. 4, Theorem 3.1.5,

2.1.1 Proposition. *The Leibniz function is*

$$\Upsilon_{BV} = \det(I - B^\dagger V)^{-t}$$

with $\tilde{V}(B,V) = V(I - B^\dagger V)^{-1}$.

Proof: We will check the cocycle identity

$$\Upsilon_{B,U+V} = \Upsilon_{UB} \Upsilon_{\tilde{V}(U,B)V}$$

That is,

$$\begin{aligned} \det\left(I - B^\dagger(U+V)\right) &= \det(I - U^\dagger B) \det\left(I - (I - B^\dagger U)^{-1} B^\dagger V\right) \\ &= \det(I - B^\dagger U) \det\left(I - (I - B^\dagger U)^{-1} B^\dagger V\right) \\ &= \det(I - B^\dagger U - B^\dagger V) \end{aligned}$$

as required. ∎

For $q > 1$, the basis $R^n \Omega$ is not orthogonal since we do not have a function of the pair products $B_{ij} V_{ij}$. However, we do have homogeneity of the degree, so that $\langle \psi_n, \psi_m \rangle = 0$ if $|n| \neq |m|$.

Now we need a basic fact about differentiating determinants. The result is readily available via the Grassmann algebra.

****

2.1.1 Grassmann algebra

We briefly review the Grassmann algebra. Starting with a vector space with basis $\{\, e_1, \ldots, e_r \,\}$, the Grassmann algebra is of dimension 2^r, with basis

$$[\![\varepsilon]\!] = \mathbf{e}_1^{\varepsilon_1} \wedge \cdots \wedge \mathbf{e}_r^{\varepsilon_r}$$

where each ε_i is 0 or 1. The multiplication rules are $\mathbf{e}_i \wedge \mathbf{e}_j = -\mathbf{e}_j \wedge \mathbf{e}_i$. In particular, $\mathbf{e}_i \wedge \mathbf{e}_i = 0$.

Given a linear mapping X on the vector space, let $\mathbf{y}_i = X_{i\lambda}\mathbf{e}_\lambda$. Then the determinant of X is the scalar satisfying

$$(\det X)\mathbf{e}_1 \wedge \cdots \wedge \mathbf{e}_r = \mathbf{y}_1 \wedge \cdots \wedge \mathbf{y}_r \qquad (2.1.1.1)$$

Now,

2.1.1.1 Proposition. *Let* $f(X_{11}, \ldots, X_{rr}) = \det X$, *for a square matrix* (X_{ij}). *Then*

$$\frac{1}{\det X} \frac{\partial}{\partial X_{ij}} \det X = (X^{-1})_{ji}$$

off the singular set $\det X = 0$.

Proof: From eq. (2.1.1.1), with $\mathbf{y}_i = X_{i\lambda}\mathbf{e}_\lambda$,

$$\frac{\partial}{\partial X_{ij}} \det X \mathbf{e}_1 \wedge \cdots \wedge \mathbf{e}_r = \mathbf{y}_1 \wedge \cdots \wedge \mathbf{e}_j \wedge \cdots \wedge \mathbf{y}_r$$

with $\mathbf{e}_j$ in the ith position. Now multiply by X_{kj} and sum over j to get

$$X_{k\mu}\frac{\partial}{\partial X_{i\mu}} (\det X)\mathbf{e}_1 \wedge \cdots \wedge \mathbf{e}_r = \mathbf{y}_1 \wedge \cdots \wedge \mathbf{y}_k \wedge \cdots \wedge \mathbf{y}_r$$

which is zero if $k \neq i$, while $k = i$ recovers $\det X$.　∎

In the section on spaces of type II, we will find representations of the Grassmann algebra — fermion algebra.

****

Now we can find the boson realization of the algebra in the induced representation. First we have a useful lemma, the proof of which is direct calculation.

2.1.2 Lemma. *For* $p \times q$ *matrices* V *and* B, *we have the identity*

$$V(I - B^\dagger V)^{-1} = V + V\left[B(I - V^\dagger B)^{-1}\right]^\dagger V$$

Now for the main result.

2.1.3 Theorem. *For the induced representation, in boson variables:*

$$R_{ij} = \mathcal{R}_{ij}, \quad L_{ij} = t\mathcal{V}_{ij} + \mathcal{R}_{\mu\lambda}\mathcal{V}_{i\lambda}\mathcal{V}_{\mu j}$$

Remark. We can write this compactly using *Wick ordering*, or *normal ordering*. I.e., any expression in $\mathcal{R}, \mathcal{V}$ with Wick dots (colons) means that it is implied that all $\mathcal{R}$'s are to the left and all $\mathcal{V}$'s are to the right, even though, as is clear, they do not commute. E.g, $:\mathcal{V}\mathcal{R}: = \mathcal{R}\mathcal{V}$, without worrying about commutation terms. Thus, the above equation for L can be written

$$L = t\mathcal{V} + :\mathcal{V}\mathcal{R}^\dagger \mathcal{V}:$$

Proof: With $\Upsilon_{BV} = \det(I - B^\dagger V)^{-t}$, taking logarithms, using Proposition 2.1.1.1, and the chain rule,

$$\frac{1}{\Upsilon}\frac{\partial\Upsilon}{\partial B_{ij}} = t(I - V^\dagger B)^{-1}_{j\mu} V_{i\mu} = t\left(V(I - B^\dagger V)^{-1}\right)_{ij}$$

$$\frac{1}{\Upsilon}\frac{\partial\Upsilon}{\partial V_{ij}} = t(I - V^\dagger B)^{-1}_{\mu j} B_{i\mu} = t\left(B(I - V^\dagger B)^{-1}\right)_{ij}$$

By Lemma 2.1.2,

$$t^{-1}\frac{1}{\Upsilon}\frac{\partial\Upsilon}{\partial B_{ij}} = V + V\left[B(I - V^\dagger B)^{-1}\right]^\dagger V$$

Therefore

$$\frac{\partial\Upsilon}{\partial B_{ij}} = tV_{ij}\Upsilon + V_{i\lambda}\frac{\partial\Upsilon}{\partial V_{\mu\lambda}}V_{\mu j}$$

On the coherent states $e^{VR}\Omega$, the partial differentiation with respect to V_{ij} is multiplication by R_{ij}, and the result follows. ∎

2.1.4 Corollary. *The element ρ_0 is given by*

$$\rho_0 = pqtI + (p + q)\mathcal{R}_{\lambda\mu}\mathcal{V}_{\lambda\mu}$$

Proof: First, calculate

$$[L_{ij}, R_{ij}] = tI + \mathcal{R}_{\mu j}\mathcal{V}_{\mu j} + \mathcal{R}_{i\lambda}\mathcal{V}_{i\lambda}$$

Then the result follows from Proposition 2.1. ∎

III. Space of skew-symmetric matrices

Now we look at the space of skew-symmetric matrices. This involves the fermion algebra and the spin representation of so($2p$).

3.1 CARTAN DECOMPOSITION

The space of $p \times p$ skew-symmetric matrices embeds in so($2p$) with the typical element

$$X = \begin{pmatrix} \alpha & v \\ \beta & \delta \end{pmatrix}$$

with all blocks $p \times p$, v and β skew-symmetric.

First a Lemma from linear algebra.

3.1.1 Lemma.   *Let U be an $r \times r$ matrix over $\mathbf{C}$ with $\operatorname{tr} U = 0$. Then:*

1. *If UV is symmetric for all symmetric V, then $U = 0$.*

2. *If UV is skew-symmetric for all skew-symmetric V, then $U = 0$.*

Proof: For #1, taking $V = diag(\mathbf{e}_i)$, i.e., with Jordan map $\mathcal{R}_i \mathcal{V}_i$, for $1 \leq i \leq r$, shows that U is diagonal. Now, multiplying by V that exchanges columns j and j', i.e., with Jordan map $\mathcal{R}_j \mathcal{V}_{j'} + \mathcal{R}_{j'} \mathcal{V}_j$, shows that $U = cI$ for some scalar c. Then $\operatorname{tr} U = 0$ yields the result.

For #2, multiplying by V that exchanges columns, flipping the sign of one of them, i.e., $\mathcal{R}_j \mathcal{V}_{j'} - \mathcal{R}_{j'} \mathcal{V}_j$, puts, up to sign, $U_{jj'}$ and $U_{j'j}$ on the diagonal, puts the diagonal entries, U_{jj} and $U_{j'j'}$ in their places instead, and zeroes out the rest of the diagonal. A 2×2 calculation is sufficient to see what is happening:

$$UV = \begin{pmatrix} a & b \\ c & d \end{pmatrix} \begin{pmatrix} 0 & 1 \\ -1 & 0 \end{pmatrix} = \begin{pmatrix} -b & a \\ -d & c \end{pmatrix}$$

which is skew only when $b = c = 0$, $a = d$. Thus, $U = aI$ and $\operatorname{tr} U = 0$ yields the result.
∎

Now we see the structure of the Lie algebra.

3.1.2 Proposition.   *A typical element of the Lie algebra, $X = \begin{pmatrix} \alpha & v \\ \beta & \delta \end{pmatrix}$, for the group acting on the space of skew-symmetric matrices, satisfies*

$$\alpha = -\delta^\dagger$$

Proof: By Theorem 1.7.2 of Ch. 4, we have $e^{VR} X e^{-VR} \Omega = (\dot{V}R + t\dot{H})\Omega$ in the induced representation. From the general forms, Ch. 4, §1.8, with $\beta \rightarrow -\beta$, we have

$$e^{VR} X e^{-VR} = \begin{pmatrix} \alpha - V\beta & v + \alpha V - V\delta - V\beta V \\ \beta & \beta V + \delta \end{pmatrix}$$

Thus,

$$\dot{V} = v + \alpha V - V\delta - V\beta V$$

must be skew-symmetric. I.e.,

$$-v - \alpha V + V\delta + V\beta V = v^\dagger + V^\dagger \alpha^\dagger - \delta^\dagger V^\dagger - V^\dagger \beta^\dagger V^\dagger$$

Given that V, v, β are all skew-symmetric yields

$$V(\alpha^\dagger + \delta) = (\alpha + \delta^\dagger)V$$

This means that $U = \alpha + \delta^\dagger$ satisfies the condition #2 of Lemma 3.1.1, including $\operatorname{tr} U = \operatorname{tr} X = 0$. Thus the result. ∎

We write X in the form

$$X = \begin{pmatrix} -\delta^\dagger & v \\ \beta & \delta \end{pmatrix}$$

The Jordan map realization in boson variables is:

$$R_{ij} = \mathcal{R}_i \mathcal{V}_{j+p} - \mathcal{R}_j \mathcal{V}_{i+p}$$
$$L_{ij} = \mathcal{R}_{i+p} \mathcal{V}_j - \mathcal{R}_{j+p} \mathcal{V}_i$$

Remark. This is defined only for $i < j$. It will be convenient later to think of R as a skew-symmetric matrix, i.e., we extend the definition of R_{ij} to $R_{ij} = -R_{ji}$ if $i > j$.

Calculate, for $i < j$,

$$[L_{ij}, R_{ij}] = \mathcal{R}_i \mathcal{V}_i + \mathcal{R}_j \mathcal{V}_j - \mathcal{R}_{i+p} \mathcal{V}_{i+p} - \mathcal{R}_{j+p} \mathcal{V}_{j+p}$$

Thus,

3.1.3 Proposition.   *The element ρ_0 satisfies*

$$(p-1)\rho_0 = \sum_{i<j} [L_{ij}, R_{ij}]$$

Up until now we have mentioned and used only the boson calculus, apart from a very brief mention of Grassmann algebra. Now we discuss fermions.

3.2 FERMIONS

The sl(2) matrices, Ch. 5. §4.1, here writing ρ instead of Λ,

$$R = \begin{pmatrix} 0 & 0 \\ 1 & 0 \end{pmatrix}, \quad L = \begin{pmatrix} 0 & 1 \\ 0 & 0 \end{pmatrix}, \quad \rho = \begin{pmatrix} 1 & 0 \\ 0 & -1 \end{pmatrix}$$

in addition to being a basis for a Lie algebra, satisfy special *anti-commutation* properties:

$$\{X, Y\} = XY + YX$$

This is a symmetric bilinear product. It is convenient to refer to the anti-commutation rules as *fermi products*. In fact, one quickly checks

3.2.1 Proposition.   *The 2×2 matrices R, L, ρ, satisfy the fermi products*

$$\begin{aligned} \{R, R\} = \{L, L\} = 0 \qquad & \{\rho, \rho\} = 2I \\ \{L, \rho\} = \{R, \rho\} = 0 \quad , \quad & \{L, R\} = I \end{aligned}$$

In general, the boson algebra is built up by raising operators $\mathcal{R}_i$ that commute. The fermion space is built up by anti-commuting raising operators: f_i^+. In fact, for given $N > 0$, they generate a Grassmann algebra. The basis is thus

$$[\![\varepsilon]\!] = (f_1^+)^{\varepsilon_1} \cdots (f_N^+)^{\varepsilon_N} \Omega$$

The annihilation operators, f_i, are adjoint to f_i^+, and satisfy $f_i \Omega = 0$, $1 \le i \le N$. The fermi products are

$$\{f_i^+, f_j^+\} = \{f_i, f_j\} = 0, \quad \{f_i, f_j^+\} = \delta_{ij} I$$

for $1 \le i, j, \le N$. These are analogous to the boson commutation relations, replacing commutators with anti-commutators.

The raising operator f_i^+ thus can be realized as a linear operator on the Grassmann algebra acting on a vector on the left as $e_i \wedge$. On the basis $[\![\varepsilon]\!]$,

$$f_j^+[\![\varepsilon]\!] = (-1)^{\varepsilon_1 + \cdots \varepsilon_{j-1}} [\![\varepsilon + \mathbf{e}_j]\!] \quad \bmod 2$$
$$f_j[\![\varepsilon]\!] = (-1)^{\varepsilon_1 + \cdots \varepsilon_{j-1}} \varepsilon_j [\![\varepsilon - \mathbf{e}_j]\!]$$

which are similar to the action of boson operators, but signs must be kept track of.

Using sl(2) matrices, one shows by induction the following matrix realization. These are, in fact, the matrices of the action on the Grassmann algebra with respect to the basis $[\![\varepsilon]\!]$ appropriately ordered. First, recall the Kronecker product of matrices (matrix corresponding to the tensor product of linear operators). In block form, for $N \times N$,

$$A \otimes B = \begin{pmatrix} a_{11}B & a_{12}B & \cdots & \cdots \\ \vdots & \vdots & \ddots & \vdots \\ a_{N1}B & \cdots & \cdots & a_{NN}B \end{pmatrix}$$

And the fermion algebra is given by

3.2.2 Proposition.   *For given $N > 0$, a matrix realization of the fermion operators is*

$$f_j^+ = I^{\otimes(N-j)} \otimes R \otimes \rho^{\otimes(j-1)}$$
$$f_j = I^{\otimes(N-j)} \otimes L \otimes \rho^{\otimes(j-1)}$$

The fermi products are checked using Proposition 3.2.1. An important feature is the property $\rho^2 = I$. Observe also that the matrices f_j and f_j^+ are in fact transposes of one another.

Examples.   For $N = 2$: $f_1^+ = I \otimes R$, $f_2^+ = R \otimes \rho$. In block form

$$f_1^+ = \begin{pmatrix} R & 0 \\ 0 & R \end{pmatrix}, \quad f_2^+ = \begin{pmatrix} 0 & 0 \\ \rho & 0 \end{pmatrix}$$

with f_1 and f_2 the corresponding transposes.

For $N = 3$, in block form, $f_1^+ = diag(R, R, R, R)$, and

$$f_2^+ = \begin{pmatrix} 0 & 0 & 0 & 0 \\ \rho & 0 & 0 & 0 \\ 0 & 0 & 0 & 0 \\ 0 & 0 & \rho & 0 \end{pmatrix}, \quad f_3^+ = \begin{pmatrix} 0 & 0 & 0 & 0 \\ 0 & 0 & 0 & 0 \\ \rho & 0 & 0 & 0 \\ 0 & -\rho & 0 & 0 \end{pmatrix}$$

Algebraically, another way to view the same construction is to think of the different factors of the tensor product as taking N mutually commuting copies of the sl(2) matrices, $\{ R_i, L_i, \rho_i \}$. Then the operators f_i^+ are built from the R's by padding with ρ's, and similarly for the f_i. E.g., for $N = 4$, we can write

$$f_1^+ = R_4, f_2^+ = R_3\rho_4, f_3^+ = R_2\rho_3\rho_4, f_4^+ = R_1\rho_2\rho_3\rho_4$$
$$f_1 = L_4, f_2 = L_3\rho_4, f_3 = L_2\rho_3\rho_4, f_4 = L_1\rho_2\rho_3\rho_4$$

In this form the fermi products are readily seen to hold.

Now, the spin representation of SO($2p$) is the group preserving the fermi products, i.e., linear automorphisms of the fermi algebra. For a block matrix $\begin{pmatrix} U & W \\ X & Y \end{pmatrix}$ to preserve fermi products means that

$$\begin{pmatrix} \phi^+ \\ \phi \end{pmatrix} = \begin{pmatrix} U & W \\ X & Y \end{pmatrix} \begin{pmatrix} f^+ \\ f \end{pmatrix}$$

satisfy the same fermi rules as do the f's. I.e.,

$$\{\phi_i, \phi_j^+\} = \{X_{i\mu}f_\mu^+ + Y_{i\mu}f_\mu, U_{j\lambda}f_\lambda^+ + W_{j\lambda}f_\lambda\} = \delta_{ij}I$$
$$\{\phi_i^+, \phi_j^+\} = \{U_{i\mu}f_\mu^+ + W_{i\mu}f_\mu, U_{j\lambda}f_\lambda^+ + W_{j\lambda}f_\lambda\} = 0$$

and similarly for the ϕ_i. From these equations follows

3.2.3 Proposition. *The matrix* $\begin{pmatrix} U & W \\ X & Y \end{pmatrix}$ *preserves fermi products if and only if:*

1. $UW^\dagger$ *and* $XY^\dagger$ *are skew-symmetric.*
2. $YU^\dagger + XW^\dagger = I$

Now we have the main fact.

3.2.4 Theorem. *The group* **G** *with Lie algebra* $\{ \begin{pmatrix} -\delta^\dagger & v \\ \beta & \delta \end{pmatrix} \}$*, with* v *and* β *skew-symmetric, satisfies the conditions of Proposition 3.2.3.*

Proof: The group element is of the form

$$g = \begin{pmatrix} E + VDB & VD \\ DB & D \end{pmatrix} = \begin{pmatrix} U & W \\ X & Y \end{pmatrix}$$

Since

$$\begin{pmatrix} E & 0 \\ 0 & D \end{pmatrix} = \exp\left[\begin{pmatrix} -\delta^\dagger & 0 \\ 0 & \delta \end{pmatrix} \right]$$

it follows that $E = (D^{-1})^\dagger$. Now calculate

$$UW^\dagger = (E + VDB)D^\dagger V^\dagger = V^\dagger + VDBD^\dagger V^\dagger$$
$$XY^\dagger = DBD^\dagger$$
$$YU^\dagger = D(E^\dagger + B^\dagger D^\dagger V^\dagger) = I + DB^\dagger D^\dagger V^\dagger$$
$$XW^\dagger = DBD^\dagger V^\dagger$$

Since B and V are skew-symmetric, property #1 holds. The second property follows from the skew-symmetry of B. ∎

3.3 COHERENT STATES

Generally, X has the form

$$X = \begin{pmatrix} -\delta^\dagger & v \\ \beta & \delta \end{pmatrix}$$

Now, for ρ_0, $\delta = -I$, while for $X \neq \rho_0$, $\text{tr}\,\delta = 0$, so that the determinant of the lower right-hand corner of the corresponding group element equals one. For the induced representation, we take $\rho_0 \Omega = pt\Omega$. Since

$$\det(\text{lower right corner } e^{\rho_0 H_0}) = e^{-pH_0}$$

the relation $e^{H_0} = (\det D)^{-1/p}$ holds. Then the group action $g\Omega = e^{tH} e^{VR} \Omega$ yields the Leibniz function

$$\Upsilon_{BV} = \det(I + BV)^{-t}$$

with $\tilde{V}(B, V) = V(I + BV)^{-1}$. This is the same as that for $su(p, p)$ restricted to skew-symmetric B, so the proof of the cocycle identity goes through.

3.3.1 Theorem.   *In terms of boson variables, for the induced representation,
for $1 \leq i < j \leq p$,*

$$R_{ij} = \mathcal{R}_{ij}, \quad L_{ij} = 2t\mathcal{V}_{ij} - \mathcal{R}_{\lambda\mu}\mathcal{V}_{i\lambda}\mathcal{V}_{\mu j}$$

*where in the summations, the anti-symmetry conditions $\mathcal{R}_{ij} = -\mathcal{R}_{ji}$, $\mathcal{V}_{ij} = -\mathcal{V}_{ji}$ for $i > j$
are used. Thus, in compact form:*

$$L = 2t\mathcal{V} - :\mathcal{V}\mathcal{R}\mathcal{V}:$$

Proof: Consider $f(B) = \det(I + BV)^{-t}$. Then, for skew-symmetric B and V:

$$\Upsilon_{BV} = f(0, B_{12}, \ldots, B_{1p}, -B_{12}, 0, B_{23}, \ldots, 0)$$

Thus, for $i < j$,

$$\frac{1}{\Upsilon} \frac{\partial \Upsilon}{\partial B_{ij}} = \frac{1}{f}\left(\frac{\partial f}{\partial B_{ij}} - \frac{\partial f}{\partial B_{ji}}\right)$$

and similarly for differentiation with respect to V_{ij}.

Now, we have seen that $f^{-1}\partial f/\partial B_{ij} = t(V(I + BV)^{-1})_{ij}$, so that

$$\frac{1}{\Upsilon}\frac{\partial \Upsilon}{\partial B_{ij}} = t\left[\left(V(I+BV)^{-1}\right)_{ij} - \left(V(I+BV)^{-1}\right)_{ji}\right]$$

$$\frac{1}{\Upsilon}\frac{\partial \Upsilon}{\partial V_{ij}} = t\left[\left(B(I+VB)^{-1}\right)_{ij} - \left(B(I+VB)^{-1}\right)_{ji}\right]$$

As in the proof of Theorem 2.1.3, convert the terms in $\partial\Upsilon/\partial B_{ij}$ using Lemma 2.1.2. Here
we need the transposed version as well:

$$\left[V(I+BV)^{-1}\right]^\dagger = V^\dagger + V^\dagger\left[B(I+VB)^{-1}\right]V^\dagger$$
$$= -V + V\left[B(I+VB)^{-1}\right]V$$

Substituting into the equation for $\partial\Upsilon/\partial B_{ij}$, subtracting the transposed relation from the
original form of Lemma 2.1.2, yields

$$\frac{\partial\Upsilon}{\partial B_{ij}} = 2tV_{ij} - V_{i\lambda}\frac{\partial\Upsilon}{\partial V_{\lambda\mu}}V_{\mu j}$$

and hence the result upon replacing the derivatives with respect to the V's by the corresponding R's.   ∎

Consequently, for $i < j$,

$$[L_{ij}, R_{ij}] = 2tI - \mathcal{R}_{j\mu}\mathcal{V}_{\mu j}\mathcal{R}_{\lambda i}\mathcal{V}_{i\lambda}$$
$$= 2tI + \mathcal{R}_{j\mu}\mathcal{V}_{j\mu} + \mathcal{R}_{i\lambda}\mathcal{V}_{i\lambda}$$

Thus,

$$\sum_{i<j}[L_{ij}, R_{ij}] = p(p-1)t + (p-1)\mathcal{R}_{\lambda\mu}\mathcal{V}_{\lambda\mu}$$

and

3.3.2 Proposition.   *In the induced representation,*

$$\rho_0 = ptI + 2\mathcal{R}_{\lambda\mu}\mathcal{V}_{\lambda\mu}\theta_{\lambda\mu}$$

Proof: From Proposition 3.1.3, the factor $p-1$ drops out. The term $\mathcal{R}_{\lambda\mu}\mathcal{V}_{\lambda\mu}$ converts to $2\mathcal{R}_{\lambda\mu}\mathcal{V}_{\lambda\mu}\theta_{\lambda\mu}$ using anti-symmetry of both factors for $\lambda > \mu$ and reducing all terms of the sum to subscripts with $\lambda < \mu$.   ∎

Notice that this last term acts indeed as twice the number operator, as it should according to the commutation rules $[\rho_0, R] = 2R$ and $[L, \rho_0] = 2L$.

IV.   Space of symmetric matrices

The group acting on the space of symmetric matrices is $\mathrm{Sp}(2p)$, the group preserving an alternating bilinear form. Here we will see this as preserving the boson commutation relations.

4.1 CARTAN DECOMPOSITION

A typical element of the Lie algebra is of the form

$$X = \begin{pmatrix} -\delta^\dagger & v \\ -\beta & \delta \end{pmatrix}$$

with v and β symmetric $p \times p$ blocks. And $\mathrm{tr}\,\delta = 0$ for $X \neq \rho_0$, while $\rho_0 = \begin{pmatrix} I & 0 \\ 0 & -I \end{pmatrix}$. The form of X follows as in Proposition 3.1.2, using #1 of Lemma 3.1.1, with the proof modified for symmetric v, β, V.

From the matrix form, the Jordan map gives, for $i \leq j$,

$$R_{ij} = \mathcal{R}_i\mathcal{V}_{j+p} + \mathcal{R}_j\mathcal{V}_{i+p}$$
$$L_{ij} = -\mathcal{R}_{i+p}\mathcal{V}_j - \mathcal{R}_{j+p}\mathcal{V}_i$$

with

$$[L_{ij}, R_{ij}] = \mathcal{R}_i\mathcal{V}_i + \mathcal{R}_j\mathcal{V}_j - \mathcal{R}_{i+p}\mathcal{V}_{i+p} - \mathcal{R}_{j+p}\mathcal{V}_{j+p}$$

so that

$$\sum_{i \leq j}[L_{ij}, R_{ij}] = p\rho_0 \tag{4.1.1}$$

Now we will see that in this case we have the symplectic group preserving boson commutation relations analogous to the orthogonal group for the case of skew-symmetric matrices preserving the fermion anti-commutation relations.

4.2 SYMPLECTIC GROUP

The spin representation of $Sp(2p)$ is the group of linear automorphisms of the boson algebra. For block $\begin{pmatrix} U & W \\ X & Y \end{pmatrix}$ to preserve boson products means that

$$\begin{pmatrix} \mathcal{R}' \\ \mathcal{V}' \end{pmatrix} = \begin{pmatrix} U & W \\ X & Y \end{pmatrix} \begin{pmatrix} \mathcal{R} \\ \mathcal{V} \end{pmatrix}$$

satisfy

$$[\mathcal{V}_i', \mathcal{V}_j'] = [\mathcal{R}_i', \mathcal{R}_j'] = 0, \ [\mathcal{V}_i', \mathcal{R}_j'] = \delta_{ij} I$$

I.e.,

$$[\mathcal{V}_i', \mathcal{V}_j'] = [X_{i\mu}\mathcal{R}_\mu + Y_{i\mu}\mathcal{V}_\mu, X_{j\lambda}\mathcal{R}_\lambda + Y_{j\lambda}\mathcal{V}_\lambda] = 0$$
$$[\mathcal{V}_i', \mathcal{R}_j'] = [X_{i\mu}\mathcal{R}_\mu + Y_{i\mu}\mathcal{V}_\mu, U_{j\lambda}\mathcal{R}_\lambda + W_{j\lambda}\mathcal{V}_\lambda] = \delta_{ij} I$$

and similarly for $[\mathcal{R}_i', \mathcal{R}_j']$. Thus, we read off from these equations

4.2.1 Proposition. *The block matrix $\begin{pmatrix} U & W \\ X & Y \end{pmatrix}$ preserves boson products if and only if*

1. $YX^\dagger$ and $WU^\dagger$ are symmetric.

2. $YU^\dagger - XW^\dagger = I$.

Thus,

4.2.2 Theorem. *The group $\mathbf{G}$ with Lie algebra $\begin{pmatrix} -\delta^\dagger & v \\ -\beta & \delta \end{pmatrix}$ with v and β symmetric, satisfies the conditions of Proposition 4.2.1.*

Proof: As in Theorem 3.2.4, the group element is of the form

$$\begin{pmatrix} U & W \\ X & Y \end{pmatrix} = \begin{pmatrix} E - VDB & VD \\ -DB & D \end{pmatrix}$$

with $E = (D^{-1})^\dagger$. Here:

$$YX^\dagger = -DB^\dagger D^\dagger$$
$$WU^\dagger = VD(E^\dagger - B^\dagger D^\dagger V^\dagger) = V - VDB^\dagger D^\dagger V^\dagger$$
$$YU^\dagger = D(E^\dagger - B^\dagger D^\dagger V^\dagger) = I - DB^\dagger D^\dagger V^\dagger$$
$$XW^\dagger = -DBD^\dagger V^\dagger$$

Now property #1 follows from the symmetry of B and V. Property #2 follows from the symmetry of B. ∎

4.3 COHERENT STATES

As in §3.3, we take for the induced representation: $\rho_0 = pt\Omega$. The Leibniz function is

$$\Upsilon_{BV} = \det(I - BV)^{-t}$$

with $\tilde{V}(B,V) = V(I - BV)^{-1}$, the same as for $su(p,p)$ with symmetric B, so that the cocycle identity follows. Now,

4.3.1 Theorem. *In terms of boson variables, for the induced representation, for $i \leq j$,*

$$\mathcal{R}_{ij} = \mathcal{R}_{ij}$$

$$L_{ij} = \left(1 - \frac{\delta_{ij}}{2}\right)\left[2t\mathcal{V}_{ij} + \mathcal{R}_{\lambda\mu}\mathcal{V}_{i\lambda}\mathcal{V}_{j\mu} + \mathcal{R}_{\lambda\lambda}\mathcal{V}_{i\lambda}\mathcal{V}_{j\lambda}\right]$$

where in summing over indices $i > j$, the terms $\mathcal{R}_{ij}$ and $\mathcal{V}_{ij}$ are replaced by $\mathcal{R}_{ji}, \mathcal{V}_{ji}$ respectively.

Proof: As in the proof of Theorem 3.3.1, for $f(B) = \det(I - BV)^{-t}$, $\Upsilon_{BV} = f(B_{11}, B_{12}, \ldots, B_{1p}, B_{12}, B_{22}, B_{23}, \ldots, B_{pp})$, so that for $i \leq j$,

$$\frac{1}{\Upsilon}\frac{\partial\Upsilon}{\partial B_{ij}} = \frac{1}{f}\left(\frac{\partial f}{\partial B_{ij}} + \frac{\partial f}{\partial B_{ji}}\right)\left(1 - \frac{\delta_{ij}}{2}\right)$$

Thus, via $f^{-1}\partial f/\partial B_{ij} = t(V(I - BV)^{-1})_{ij}$ and similarly for differentiation with respect to V_{ij},

$$\frac{1}{\Upsilon}\frac{\partial\Upsilon}{\partial B_{ij}} = t\left[(V(I - BV)^{-1})_{ij} + (V(I - BV)^{-1})_{ji}\right]\left(1 - \frac{\delta_{ij}}{2}\right)$$

$$\frac{1}{\Upsilon}\frac{\partial\Upsilon}{\partial V_{ij}} = t\left[(B(I - VB)^{-1})_{ij} + (B(I - VB)^{-1})_{ji}\right]\left(1 - \frac{\delta_{ij}}{2}\right)$$

For symmetric B and V:

$$V(I - BV)^{-1} = V + V\left[B(I - VB)^{-1}\right]^\dagger V$$

$$\left[V(I - BV)^{-1}\right]^\dagger = V + VB(I - VB)^{-1}V \tag{4.3.1}$$

Write $C = (I - VB)^{-1}$. Then, calling r the array with entries $r_{ij} = \Upsilon^{-1}\partial\Upsilon/\partial V_{ij}$, denoting the diagonal submatrix of the product BC by $\mathrm{diagl}(BC)$,

$$t\left(BC + (BC)^\dagger\right) = r + \mathrm{diagl}\,(BC)$$

But, $\Upsilon^{-1}\partial\Upsilon/\partial V_{ii} = t(BC)_{ii}$, i.e., $\mathrm{diagl}(BC)$ corresponds exactly to $\mathcal{R}_{11}, \ldots, \mathcal{R}_{pp}$. Now, adding equations (4.3.1), with the overall factor of $\left(1 - \frac{\delta_{ij}}{2}\right)$, the result follows. ∎

Now, for $i < j$,

$$[L_{ij}, R_{ij}] = 2tI + \mathcal{R}_{j\mu}\mathcal{V}_{j\mu} + \mathcal{R}_{\lambda i}\mathcal{V}_{i\lambda} + \mathcal{R}_{jj}\mathcal{V}_{jj} + \mathcal{R}_{ii}\mathcal{V}_{ii}$$

For $i = j$,

$$[L_{ii}, R_{ii}] = tI + \mathcal{R}_{i\lambda}\mathcal{V}_{i\lambda} + \mathcal{R}_{ii}\mathcal{V}_{ii}$$

Summing,

$$\sum_{i \leq j}[L_{ij}, R_{ij}] = ptI + \mathcal{R}_{\lambda\mu}\mathcal{V}_{\lambda\mu} + \mathcal{R}_{\lambda\lambda}\mathcal{V}_{\lambda\lambda}$$

$$+ (p^2 - p)tI + (p - 1)(\mathcal{R}_{\lambda\mu}\mathcal{V}_{\lambda\mu} + \mathcal{R}_{\lambda\lambda}\mathcal{V}_{\lambda\lambda})$$

$$= p^2 tI + p(\mathcal{R}_{\lambda\mu}\mathcal{V}_{\lambda\mu} + \mathcal{R}_{\lambda\lambda}\mathcal{V}_{\lambda\lambda})$$

Hence, from eq. (4.1.1), putting subscripts in increasing order,

$$\rho_0 = ptI + 2\mathcal{R}_{\lambda\mu}\mathcal{V}_{\lambda\mu}\theta_{\lambda\mu} + 2\mathcal{R}_{\lambda\lambda}\mathcal{V}_{\lambda\lambda}$$

which indeed acts as twice the number operator on $\mathcal{R}_{ij}$ and $\mathcal{V}_{ij}$ for $i \leq j$.

4.4 CANONICAL APPELL SYSTEMS

Here, starting from R_{ij}, we have a commuting family of self-adjoint operators constructed by the adjoint action of the group. Let $Y = \sum_i L_{ii}$. Then, in block matrix form,

$$Y = \begin{pmatrix} 0 & 0 \\ -I & 0 \end{pmatrix} \text{ and}$$

$$e^Y vRe^{-Y} = \begin{pmatrix} I & 0 \\ -I & I \end{pmatrix}\begin{pmatrix} 0 & v \\ 0 & 0 \end{pmatrix}\begin{pmatrix} I & 0 \\ I & I \end{pmatrix} = \begin{pmatrix} v & v \\ -v & -v \end{pmatrix}$$

while its adjoint is

$$\exp(-\sum_i R_{ii}) vL \exp(\sum_i R_{ii}) = \begin{pmatrix} I & -I \\ 0 & I \end{pmatrix}\begin{pmatrix} 0 & 0 \\ -v & 0 \end{pmatrix}\begin{pmatrix} I & I \\ 0 & I \end{pmatrix} = \begin{pmatrix} v & v \\ -v & -v \end{pmatrix}$$

Thus, $X_{ij} = e^Y R_{ij}e^{-Y}$ yields a commuting family of self-adjoint operators. We can find the joint distribution:

$$\langle e^{z_{\lambda\mu}X_{\lambda\mu}} \rangle = \langle e^Y e^{zR} e^{-Y} \rangle$$

$$= \Upsilon_{Iz} = \det(I - z)^{-t}$$

This is the *Wishart distribution* on the space of symmetric matrices.

References

The works of Hua [35], Satake [48], and Wolf [55] are recommended for the study of symmetric domains. Jakobsen [37] studies unitarity of representations associated to these domains.

The Hermitian symmetric spaces and more generally Kähler manifolds admit quantization. This was developed by Berezin [2]. See Perelomov [47] as well.

Chapter 8   Properties of matrix elements

In this brief chapter, we put together the basic special function properties of the theory. Starting from the right dual, the principal formula produces the matrix elements. They satisfy an addition formula according to the group law. They satisfy recurrence relations as well, which we will find in section two of this chapter. For quotient representations, we have previously seen a generating function for the matrix elements that can be found via the group law, and in section three we give a summation formula expressing the matrix elements for the quotient representation in terms of the general matrix elements.

I.   Addition formulas

Let us look at addition formulas in general. They are based on the group law:

1. Along a one-parameter subgroup, $e^{sX} = g(A(s))$, the coordinates satisfy

$$g(A(s + s')) = g(A(s))g(A(s')) = g(A(s) \odot A(s'))$$

Thus, the addition formula for the A-coordinates as functions of the parameter s:

$$A(s + s') = A(s) \odot A(s') = A(s') \odot A(s)$$

The interesting feature is the symmetry in s, s' even though the group law $A \odot A'$, for non-abelian groups, is not symmetric in A, A'. In the context of the Lie response, $A(s)$ is the response to the linear signal $\alpha(s) = s\alpha$.

2. The main addition formulas for the matrix elements are direct expressions of the fact that they are a representation of the group. Recall eqs. (3.1.2), (3.1.3) of Ch. 2:

$$\left\langle {m \atop n} \right\rangle_{A \odot A'} = \left\langle {m \atop \lambda} \right\rangle_A \left\langle {\lambda \atop n} \right\rangle_{A'}$$

$$c_m(A \odot A') = \left\langle {m \atop \lambda} \right\rangle_A c_\lambda(A')$$

For example, Proposition 3.1.2 of Ch. 2 says that the right dual pi-matrix is given by

$$\pi_{ij}^* = \left\langle {e_j \atop e_i} \right\rangle$$

Therefore,

$$\pi_{ij}^*(A \odot A') = \left\langle {e_j \atop \lambda} \right\rangle_A \left\langle {\lambda \atop e_i} \right\rangle_{A'} \tag{1.1}$$

By the principal formula,

$$\left\langle {e_j \atop \lambda} \right\rangle_A = (\xi^*)^\lambda A_j$$

Thus,

1.1 Proposition.   *If the right dual representation satisfies* $(\xi^*)^n A_j = 0$, *for* $|n| > 1$, *then the matrices* π^* *are a representation of the group.*

Proof: With this condition, the summation in eq. (1.1) reduces to a sum over single indices.   ∎

We have the examples from nilpotent groups, cf. the Heisenberg, ABCD, and upper-triangular algebras.

II. Recurrences

Now we find recurrence relations for the matrix elements of the group acting on $\mathcal{U}(\mathcal{G})$.

2.1 Lemma.   *The matrix elements* $\left\langle\begin{smallmatrix} m \\ n \end{smallmatrix}\right\rangle$ *satisfy*

$$\check{\pi}_{j\lambda}^{\dagger}(A)\pi_{\lambda\mu}^{\ddagger}(\bar{\mathcal{R}})\bar{\mathcal{V}}_{\mu}\left\langle\begin{matrix} \mathbf{m} \\ n \end{matrix}\right\rangle = \mathcal{R}_{\mu}\hat{\pi}_{\mu j}(\mathcal{V})\left\langle\begin{matrix} m \\ \mathbf{n} \end{matrix}\right\rangle$$

where the boldface indicates the index acted upon.

Proof: From eq. (2.7.2) of Ch. 2 applied to $\left\langle\begin{smallmatrix} m \\ n \end{smallmatrix}\right\rangle$:

$$\xi^*\left\langle\begin{matrix} m \\ n \end{matrix}\right\rangle = \check{\pi}^{\dagger}\xi^{\ddagger}\left\langle\begin{matrix} m \\ n \end{matrix}\right\rangle = \hat{\xi}\left\langle\begin{matrix} m \\ \mathbf{n} \end{matrix}\right\rangle = \mathcal{R}\hat{\pi}(\mathcal{V})\left\langle\begin{matrix} m \\ \mathbf{n} \end{matrix}\right\rangle$$

by the differential recurrence relations, Ch. 2, Theorem 3.2.1 and the definition of the double dual. To get the left-hand side, in the above equations, replace $\xi^{\ddagger}$ according to Ch. 2, Theorem 3.2.1 and Corollary 3.2.2, to the effect that the action of $\xi^{\ddagger}$ is given in terms of $\bar{\mathcal{R}}, \bar{\mathcal{V}}$ on the index m.   ∎

From this,

2.2 Theorem.   *The matrix elements* $\left\langle\begin{smallmatrix} m \\ n \end{smallmatrix}\right\rangle$ *satisfy the recurrence*

$$\mathcal{R}\left\langle\begin{matrix} m \\ \mathbf{n} \end{matrix}\right\rangle = (\pi^{\ddagger}(\mathcal{V}))^{-1}\check{\pi}^{\dagger}(A)\pi^{\ddagger}(\bar{\mathcal{R}})\bar{\mathcal{V}}\left\langle\begin{matrix} m \\ n \end{matrix}\right\rangle$$

where, on the right-hand side, $\bar{\mathcal{R}}, \bar{\mathcal{V}}$ *act on the index* m *and* $\mathcal{V}$ *affects only the index* n.

Remark. The arrays $\mathcal{R}$ and $\bar{\mathcal{V}}$ are taken as column vectors and the multiplication is the usual product of matrices.

Proof: In the Lemma, write $\hat{\pi}(\mathcal{V})$ as the transpose of $\pi^{\dagger}(\mathcal{V})$. Since the action of $\mathcal{V}$ on the index n commutes with the operators on the left-hand side, we can multiply through by the inverse matrix, leaving on the right-hand side the action of $\mathcal{R}$ on n. ∎

The interesting feature is that the raising operators $\mathcal{R}$ commute. In the principal formula, the n indices correspond to powers of the right dual operators ξ^*, which obey the commutation relations of $\mathcal{G}$. Thus, this is, in some sense, an *abelianization* of the Lie algebra.

This formula is, in fact, the one used in the `matrec` program. See the next chapter for discussion regarding recursive application of this general recurrence formula.

III.   Quotient representations and summation formulas

For induced representations, Proposition 4.2.1.3 of Ch. 2 gives the generating function for the matrix elements $\left\langle\!\!\left\langle \begin{smallmatrix} m \\ n \end{smallmatrix} \right\rangle\!\!\right\rangle$

$$e^{\tau H''} c_m(V'') = c_\lambda(V') \left\langle\!\!\left\langle \begin{smallmatrix} m \\ \lambda \end{smallmatrix} \right\rangle\!\!\right\rangle$$

where

$$g(V, A, B)g(V', 0, 0) = g(V'', A'', B'')$$

Now, for symmetric Lie algebras, Proposition 1.4.1 of Ch. 4 gives for the left-hand side

$$e^{\tau(H+\tilde{H}(B,V'))} c_m\left(V + e^{A\hat{\rho}} \tilde{V}(B, V')\right)$$

We can express this in terms of the Leibniz function as

$$e^{\tau H} \Upsilon_{BV'} c_m\left(V + e^{A\hat{\rho}} \tilde{V}(B, V')\right)$$

In the matrix formulation, for the symmetric spaces of Ch. 7, we thus have the generating function, e.g., cf. general forms Ch. 4, §1.8, and Proposition 2.1.1 of Ch. 7,

$$e^{tH} \det(I - B^{\dagger}V')^{-t} c_m\left(V + EV'(I - B^{\dagger}V')^{-1}D^{-1}\right)$$

a similar formula holding for type II and III spaces, with $D^{-1} = E^{\dagger}$.

3.1   SUMMATION THEOREM

For the quotient representation we have a summation theorem expressing the matrix elements $\left\langle\!\!\left\langle \begin{smallmatrix} m \\ n \end{smallmatrix} \right\rangle\!\!\right\rangle$ in terms of the $\left\langle \begin{smallmatrix} m \\ n \end{smallmatrix} \right\rangle$.

3.1.1 Proposition.   *The summation formula:*

$$\tau^{n'} \left\langle\!\left\langle \begin{matrix} m \\ n \end{matrix} \right\rangle\!\right\rangle = \sum_{m'} \left\langle \begin{matrix} m\,m'\,0\,0 \\ n\,n'\,0\,0 \end{matrix} \right\rangle \tau^{m'}$$

The indices are indicated corresponding to the decomposition $\mathcal{P} \oplus \mathcal{H} \oplus \mathcal{N} \oplus \mathcal{L}$.

Proof: Write the basis for $\mathcal{U}(\mathcal{G})$ as $R^n \rho^{n'} \rho'^{n''} L^{n'''}$, with ρ' denoting the ρ_{ij}'s, i.e., the basis for $\mathcal{N}$. Then

$$g\,R^n \rho^{n'} \rho'^{n''} L^{n'''} \Omega = \sum \left\langle \begin{matrix} m\,m'\,m''\,m''' \\ n\,n'\,n''\,n''' \end{matrix} \right\rangle R^m \rho^{m'} \rho'^{m''} L^{m'''} \Omega$$

$$= \sum \left\langle \begin{matrix} m\,m'\,0\,0 \\ n\,n'\,0\,0 \end{matrix} \right\rangle \tau^{m'} R^m \Omega$$

While $R^n \rho^{n'} \rho'^{n''} L^{n'''} \Omega = \tau^{n'} R^n \Omega$ and the action of g gives the matrix elements $\left\langle\!\left\langle \begin{matrix} m \\ n \end{matrix} \right\rangle\!\right\rangle$. Hence the result.   ∎

References

See the forthcoming article [17], which includes results for su(3). Some examples of matrix elements for quotient representations and summation formulas are given in [24].

Chapter 9   Symbolic computations

For the main ingredients of the theory, there are two basic situations:

1) a typical element $X \in \mathcal{G}$ is given as a matrix, with the basis implicit

2) $\mathcal{G}$ is described in terms of commutation relations, with the basis thus given abstractly

For 1) we outline the computations to be done:

Given X in matrix form

1. The element $X = \sum \alpha_i \xi_i$ is given as a matrix. The basis is found by $\xi_i = \dfrac{\partial X}{\partial \alpha_i}$.

2. Compute g as the product $e^{A_1 \xi_1} e^{A_2 \xi_2} \dots e^{A_d \xi_d}$.

3. Compute Xg by matrix multiplication.

4. Write $\dot{g}$ formally and equate entries of $\dot{g}$ and Xg.

5. Solve for $\dot{A}_i$. Form the row vector $\dot{A}$. Note: no differential equations are solved at this point.

6. Find $\pi^{\dagger}$ from the equation $\dot{A} = \alpha \pi^{\dagger}$, with α the row vector $(\alpha_1, \dots, \alpha_d)$.

7. Repeat steps 3–6 with gX to find π^*.

8. Form the vector fields $\xi^{\dagger} = \pi^{\dagger} \partial$, $\xi^* = \pi^* \partial$, using column vectors.

9. Form double duals $\hat{\xi}$, $\hat{\xi}^*$, using $\hat{\pi} = (\pi^{\dagger})^{\dagger}$, $\hat{\pi}^* = (\pi^*)^{\dagger}$.

10. Find the adjoint representation of the algebra as the linearization of $\hat{\pi}^* - \hat{\pi}$.

11. Compute the adjoint representation of the group $\tilde{\pi} = \hat{\pi}^{-1} \hat{\pi}^*$.

12. Find the group law.

13. Find induced representations from the double dual.

14. Solve $\dot{A} = \alpha \pi$, using either $\pi^{\dagger}$ or π^*, to find the splitting formula.

Remark. Note that using matrices, there is the possibility of computing the exponential of X directly. Equating to g yields the splitting formula directly.

For situation 2) here is an outline:

Given $\mathcal{G}$ in terms of commutation relations

1. The basis $\{ \xi_i \}$ is given, along with commutation relations. Form $X = \sum \alpha_i \xi_i$.

2. Write g as the product $e^{A_1 \xi_1} e^{A_2 \xi_2} \dots e^{A_d \xi_d}$.

3. For each ξ_i individually, compute the multiplication

$$\xi_i g = \ldots + e^{A_1 \xi_1} \cdots \xi_i e^{A_i \xi_i} \cdots$$

using the adjoint action $\xi_i e^{A_j \xi_j} = e^{A_j \xi_j} (e^{-A_j \operatorname{ad} \xi_j} \xi_i)$, rewriting each term of the form $e^{A_1 \xi_1} \cdots \xi_j e^{A_j \xi_j} \cdots$ as $\partial_j g$, with $\partial_j = \partial / \partial A_j$.

4. This gives the vector fields $\xi_i^{\dagger} = \pi_{i\mu}^{\dagger} \partial_\mu$.

5. Extract $\pi^{\dagger}$ from the relation $\xi^{\dagger} = \pi^{\dagger} \partial$ where $\xi^{\dagger}$ and ∂ are column vectors.

6. Repeat steps 3–5 for right multiplication $g\xi_i$ to find ξ^* and hence π^*.

7. Continue with double duals and adjoint representation.

8. Use either π^* or $\pi^{\dagger}$ to form the system $\dot{A} = \alpha \pi$. Solve to find the splitting formula.

For canonical Appell systems corresponding to a Cartan decomposition $\mathcal{P} \oplus \mathcal{K} \oplus \mathcal{L}$, here is an outline:

For canonical Appell systems

1. Compose the Lie algebra with basis R_i, ρ_i, ρ_{ij}, L_i.

2. Write $X = \sum v_i R_i + \sum h_l \rho_l + \sum \kappa_{ij} \rho_{ij} + \sum \beta_k L_k$.

3. Use either the matrices or commutation relations to find the adjoint orbit $e^{-VR} X e^{VR}$.

4. With $\rho_{ij}\Omega = L_k \Omega = 0$, $\rho_i \Omega = \tau_i \Omega$, equate $e^{-VR} X e^{VR} \Omega = (\tau \dot{H} + \dot{V} R)\Omega$.

5. Extract equations for $\dot{H}$ and $\dot{V}$. Solve. A possibility is to do this for simple cases, and then use the group law to generate solutions.

6. Write $zX = \sum z_j X_j$, with commuting X_j. Form the generating function $e^{zX - \tau H(z)} \Omega = e^{V(z)R} \Omega$.

7. Invert $V(z)$ to find the generating function for the basis: $e^{vR} = e^{XU(v) - \tau M(v)}$.

8. Find $H(z)$, $V(z)$ and the corresponding operator calculus.

9. Find a Leibniz function. Construct the associated Hilbert space.

10. For self-adjoint X_j, find the joint distribution from $\langle e^{zX} \rangle$.

I. Computing the pi-matrices

Now, consider the basic MAPLE procedures we have been using. First, let us remark that the procedures can conveniently be saved as ascii files and read directly into the worksheet with a read 'filename' statement. Similarly, variables can be saved to files and read in as needed to save memory. See the output for so(6), for example.

Start with the procedure lie1. First the matrix X is given (note that the variables are actually a's instead of the α's used in the theoretical discussion). The dimension of the Lie algebra is d and the size of the matrix X is $n \times n$. The procedure is called through lie in the form lie(X,d,n,ε), where ε is 0 or 1 depending on whether exponential or trigonometric form of the output is desired (e.g., for so(3), trigonometric form of the output is natural). The main computations are done using exponential form.

First, the basis matrices ξ_i are found by differentiation: $\partial X/\partial a_i$. Then the group element is calculated as a product of one-parameter subgroups generated by the ξ_i. For the left dual, Xg and gdot are found. Note that gdot is the sum of each $\partial g/\partial A_i$ times the formal variable Adot[i]. These are converted to vectors, equated, and the Adot's solved for. Then the pi-matrix is found from the splitting lemma using the equation

$$\pi^\dagger_{ij} = \frac{\partial \text{Adot}\,[j]}{\partial a_i}$$

(In the actual procedure, the vectors are set up as rows, so that in fact the transposes are constructed first.) The right dual is found similarly. The exponential of the adjoint representation is computed as $\hat{\pi}^{-1}\hat{\pi}^*$. The auxiliary procedures see and trg respectively print out the matrices and, if called for, convert output to trigonometric form.

Some general procedures come next. Calling cmm(X,Y) will produce the commutator $[X,Y]$. The procedure kll, "Killing," calculates the trace of the product of two matrices. Note that directly this is not the same as the standard matrix inner product, which would be computed using kll($X^\dagger$,Y).

If the basis ξ_i is orthogonal with respect to the standard matrix inner product, then cffs(X,xi,d) will give the expansion coefficients of X in the basis xi[1],...,xi[d]. If you only want the basis without running lie, then liealg(X), will produce it from the generic element X of $\mathcal{G}$.

Once lie has been run, kirlv() computes the Kirillov form, which shows the commutation relations for $\mathcal{G}$. The procedure matrec uses the formula given in Ch. 8, Theorem 2.2, to find the recurrence relations for the matrix elements $\langle {}^m_n \rangle$. The output is "normal ordered," i.e., all of the VV's are considered to be to the far right, all RR's to the left.

The procedure grp computes the product of group elements, and, given that the ξ_i are an orthogonal basis, outputs the coefficients of the product expanded in the basis ξ. Then one can use grpels to solve for the group law.

The dual representations are computed by duals. The output is "normal ordered," here meaning that all δ_i's, denoting the partial derivatives ∂_i's are considered to the right of all A's. If the basis is ordered appropriately, especially for the Hermitian symmetric cases, the procedure indrep will give the induced representation in terms of boson operators via the double dual, so that duals must be run first. The input is dim $\mathcal{P}$, and the next raising operator is taken as ρ_0, then the remaining R's are zero'd out. The procedure could be modified for more general induced representations. Finally, adjrep will print out the matrices of the adjoint representation after kirlv has been called.

II. Adjoint group

The method used here for computing the exponential of the adjoint representation is the equation

$$\check{\pi} = \hat{\pi}^{-1}\hat{\pi}^*$$

For the matrices of the adjoint representation, used in the procedure kirlv, we linearize the difference $\hat{\pi}^* - \hat{\pi}$, i.e., differentiate with respect to each $A[i]$ in turn and evaluate at $A = 0$. The Kirillov form is constructed by multiplying each ξ_i by the row vector $\mathbf{x}$, which produces the ith row of the Kirillov form.

The computation of the pi-matrices using the adjoint representation of the Lie algebra directly corresponding to a flag structure has not yet been implemented. The idea would be that the entire adjoint matrices are not needed, just the restrictions to the subalgebras of the flag.

III. Recursive computation of matrix elements

The principal formula gives a recursive formula for the matrix elements $\left\langle {m \atop n} \right\rangle$. Although the ξ^* do not commute, acting on polynomials, the operators of differentiation and multiplication by A effectively are the same as abstract boson velocity and raising operators.

On the other hand, with matrec, the raising of indices n is given in terms of commuting operators. Starting from $c_m(A) = \left\langle {m \atop 0} \right\rangle$ using the recurrence relations of matrec, keep the results always as linear combinations of $\left\langle {m \atop 0} \right\rangle$. One must take care to replace

$$A_i \left\langle {m \atop 0} \right\rangle = (m_i + 1) \left\langle {m + \mathbf{e}_i \atop 0} \right\rangle$$

and maintain the level $n = 0$. In this way, starting from $[\![0]\!]$, the matrix elements are found as polynomials recursively.

IV.   Symbolic computation of Appell systems

Use the form of the Lie algebra as vector fields — using right/left duals or in terms of boson operators, the double dual, or an induced representation. Starting with $[\![n]\!]$, apply $g(W(t))$:

$$g(W(t))\,[\![n]\!] = \sum_m \left\langle \begin{matrix} m \\ n \end{matrix} \right\rangle_{W(t)} [\![m]\!]$$

Now take averages with respect to the underlying w-process:

$$\langle g(W(t))[\![n]\!]\rangle_w = \left\langle \sum_m \left\langle \begin{matrix} m \\ n \end{matrix} \right\rangle_{W(t)} [\![m]\!]\right\rangle_w$$

This gives an extension of Appell systems to noncommutative polynomial functions. The matrix elements are assumed to have been computed (as in the previous section).

Now consider the induced representation, with basis $R^n\Omega$. The R's are commuting variables, so the basis can be considered as polynomials, by writing everything in terms of y-variables, replacing $R \leftrightarrow y$. In the basis $R^n\Omega$

$$\langle g(W(t))R^n\Omega\rangle_w = \left\langle \sum_m \left\langle\!\!\left\langle \begin{matrix} m \\ n \end{matrix} \right\rangle\!\!\right\rangle_{W(t)} R^m\Omega\right\rangle_w$$

The matrix elements for the induced representation can be found using the recursive form of Ch. 2, Proposition 4.2.1.1. The generating function Ch. 2, Proposition 4.2.1.3, is available as well.

In terms of polynomial functions, this is equivalent to using the induced representation from the double dual, writing the representation of the Lie algebra as vector fields. Once the matrix elements are known, and the moments of the distribution, the Appell systems are known.

On the other hand, we have seen in Ch. 6, in the examples of the Heisenberg and affine groups, that by computing directly using symbolic differentiation/boson calculus, one can find the correlation functions via the Appell system. One is free to use any realization of the Lie algebra as vector fields or boson operators that is effective in the specific situation.

References

A condensed exposition and some examples using the approach of this volume are in the report [16]. The Amsterdam group has developed a symbolic package for Lie theory, see van Leeuwen, Cohen & Lisser [44].

MAPLE Output and Procedures

This is the finite-difference algebra.

```
> X:=matrix(2,2,[[a[2]+a[3],a[1]],[0,a[1]+a[2]]]);
```

$$X := \begin{bmatrix} a_{[2]} + a_{[3]} & a_{[1]} \\ 0 & a_{[1]} + a_{[2]} \end{bmatrix}$$

```
> d:=3:n:=2:lie(X,d,n,0);
```

group element

$$\begin{bmatrix} e^{A_{[2]}} e^{A_{[3]}} & e^{A_{[1]}} e^{A_{[2]}} - e^{A_{[2]}} \\ 0 & e^{A_{[1]}} e^{A_{[2]}} \end{bmatrix}$$

exp-adjoint

$$\begin{bmatrix} \left(e^{A_{[1]}} + e^{A_{[3]}} - 1 \right) e^{-A_{[1]}} & 0 & -\left(e^{A_{[1]}} - 1 \right) e^{-A_{[1]}} \\ -\left(e^{A_{[3]}} - 1 \right) e^{-A_{[1]}} & 1 & \left(e^{A_{[1]}} - 1 \right) e^{-A_{[1]}} \\ \left(e^{A_{[3]}} - 1 \right) e^{-A_{[1]}} & 0 & e^{-A_{[1]}} \end{bmatrix}$$

left-dual

$$\begin{bmatrix} 1 & 0 & 0 \\ 0 & 1 & 0 \\ e^{A_{[1]}} - 1 & -e^{A_{[1]}} + 1 & e^{A_{[1]}} \end{bmatrix}$$

right-dual

$$\begin{bmatrix} e^{A_{[3]}} & -e^{A_{[3]}} + 1 & e^{A_{[3]}} - 1 \\ 0 & 1 & 0 \\ 0 & 0 & 1 \end{bmatrix}$$

159

```
> matrec();
```

$$\left[VV_{[1]} - \frac{VV_{[1]}\, e^{RR_{[1]}}}{e^{A_{[1]}}} + \frac{VV_{[1]}\, e^{A_{[3]}}\, e^{RR_{[1]}}}{e^{A_{[1]}}} + \frac{VV_{[2]}\, e^{RR_{[1]}}}{e^{A_{[1]}}} \right.$$

$$\left. - \frac{VV_{[2]}\, e^{A_{[3]}}\, e^{RR_{[1]}}}{e^{A_{[1]}}} - \frac{VV_{[3]}\, e^{RR_{[1]}}}{e^{A_{[1]}}} + \frac{VV_{[3]}\, e^{A_{[3]}}\, e^{RR_{[1]}}}{e^{A_{[1]}}} \right]$$

$$\left[VV_{[2]} \right]$$

$$\left[-VV_{[1]} + \frac{VV_{[1]}\, e^{RR_{[1]}}}{e^{A_{[1]}}} - \frac{VV_{[1]}\, e^{A_{[3]}}\, e^{RR_{[1]}}}{e^{A_{[1]}}} + \frac{VV_{[1]}\, e^{A_{[3]}}\, e^{RR_{[1]}}}{e^{V_{[1]}}\, e^{A_{[1]}}} + VV_{[2]} \right.$$

$$- \frac{VV_{[2]}\, e^{RR_{[1]}}}{e^{A_{[1]}}} + \frac{VV_{[2]}\, e^{A_{[3]}}\, e^{RR_{[1]}}}{e^{A_{[1]}}} - \frac{VV_{[2]}\, e^{A_{[3]}}\, e^{RR_{[1]}}}{e^{V_{[1]}}\, e^{A_{[1]}}} + \frac{VV_{[3]}\, e^{RR_{[1]}}}{e^{A_{[1]}}}$$

$$\left. - \frac{VV_{[3]}\, e^{A_{[3]}}\, e^{RR_{[1]}}}{e^{A_{[1]}}} + \frac{VV_{[3]}\, e^{A_{[3]}}\, e^{RR_{[1]}}}{e^{V_{[1]}}\, e^{A_{[1]}}} \right]$$

```
> kirlv();
```

$$\begin{bmatrix} 0 & 0 & -x_{[1]} + x_{[2]} - x_{[3]} \\ 0 & 0 & 0 \\ x_{[1]} - x_{[2]} + x_{[3]} & 0 & 0 \end{bmatrix}$$

```
> duals();
```

$$\textit{xi-dagger}_{[1]} = \delta_{[1]}$$

$$\textit{xi-dagger}_{[2]} = \delta_{[2]}$$

$$\textit{xi-dagger}_{[3]} = \delta_{[1]}\, e^{A_{[1]}} - \delta_{[1]} - \delta_{[2]}\, e^{A_{[1]}} + \delta_{[2]} + e^{A_{[1]}}\,\delta_{[3]}$$

$$\textit{xi-star}_{[1]} = e^{A_{[3]}}\,\delta_{[1]} + \delta_{[2]} - \delta_{[2]}\, e^{A_{[3]}} - \delta_{[3]} + \delta_{[3]}\, e^{A_{[3]}}$$

$$\textit{xi-star}_{[2]} = \delta_{[2]}$$

$$\textit{xi-star}_{[3]} = \delta_{[3]}$$

$$xi\text{-}hat_{[1]} = R_{[1]}$$

$$xi\text{-}hat_{[2]} = R_{[2]}$$

$$xi\text{-}hat_{[3]} = R_{[1]} \, e^{V_{[1]}} - R_{[1]} - R_{[2]} \, e^{V_{[1]}} + R_{[2]} + e^{V_{[1]}} R_{[3]}$$

```
> indrep(1);
```

$$inducedrep_{[1]} = R_{[1]}$$

$$inducedrep_{[2]} = t$$

$$inducedrep_{[3]} = R_{[1]} \, e^{V_{[1]}} - R_{[1]} - e^{V_{[1]}} \, t + t$$

This is sl(2), the Lie algebra of 2x2 matrices of trace zero. A real form is su(2) the Lie algebra of unitary matrices. It is isomorphic to so(3).

> **X:=matrix(2,2,[[a[2],a[1]],[-a[3],-a[2]]]);**

$$X_{;} = \begin{bmatrix} a_{[2]} & a_{[1]} \\ -a_{[3]} & -a_{[2]} \end{bmatrix}$$

> **d:=3:n:=2:lie(X,d,n,0);**

group element

$$\begin{bmatrix} e^{A_{[2]}} - \dfrac{A_{[1]}A_{[3]}}{e^{A_{[2]}}} & \dfrac{A_{[1]}}{e^{A_{[2]}}} \\ -\dfrac{A_{[3]}}{e^{A_{[2]}}} & \dfrac{1}{e^{A_{[2]}}} \end{bmatrix}$$

exp-adjoint

$$\left[\left(e^{4A_{[2]}} - 2A_{[1]}A_{[3]}\,\%2 + A_{[1]}^{2}A_{[3]}^{2} \right)\%1\,,\,2A_{[1]}\left(-\%2 + A_{[1]}A_{[3]}\right)\%1\,, \right.$$

$$\left. A_{[1]}^{2}\,\%1 \right]$$

$$\left[-A_{[3]}\left(-\%2 + A_{[1]}A_{[3]}\right)\%1\,,\,\left(\%2 - 2A_{[1]}A_{[3]}\right)\%1\,,\,-A_{[1]}\,\%1 \right]$$

$$\left[\%1\,A_{[3]}^{2}\,,\,2\,\%1\,A_{[3]}\,,\,\%1 \right]$$

$$\%1 := e^{-2A_{[2]}}$$

$$\%2 := e^{2A_{[2]}}$$

left-dual

$$\begin{bmatrix} 1 & 0 & 0 \\ 2\,A_{[1]} & 1 & 0 \\ A_{[1]}^{\;2} & A_{[1]} & e^{\,2\,A_{[2]}} \end{bmatrix}$$

right-dual

$$\begin{bmatrix} e^{\,2\,A_{[2]}} & A_{[3]} & A_{[3]}^{\;2} \\ 0 & 1 & 2\,A_{[3]} \\ 0 & 0 & 1 \end{bmatrix}$$

> **grp();**

group – entries

table([

$$2 = \tfrac{1}{2}\,e^{A_{[2]}}\,e^{B_{[2]}} - \tfrac{1}{2}\,e^{A_{[2]}}\,B_{[1]}\,e^{-B_{[2]}}\,B_{[3]} - \tfrac{1}{2}\,A_{[1]}\,e^{-A_{[2]}}\,A_{[3]}\,e^{B_{[2]}}$$

$$+ \tfrac{1}{2}\,A_{[1]}\,e^{-A_{[2]}}\,A_{[3]}\,B_{[1]}\,e^{-B_{[2]}}\,B_{[3]} - \tfrac{1}{2}\,A_{[1]}\,e^{-A_{[2]}}\,e^{-B_{[2]}}\,B_{[3]}$$

$$+ \tfrac{1}{2}\,e^{-A_{[2]}}\,A_{[3]}\,B_{[1]}\,e^{-B_{[2]}} - \tfrac{1}{2}\,e^{-A_{[2]}}\,e^{-B_{[2]}}$$

$$1 = B_{[1]}\,e^{-B_{[2]}}\,e^{A_{[2]}} - B_{[1]}\,e^{-B_{[2]}}\,A_{[1]}\,e^{-A_{[2]}}\,A_{[3]} + A_{[1]}\,e^{-A_{[2]}}\,e^{-B_{[2]}}$$

$$3 = e^{-A_{[2]}}\,A_{[3]}\,e^{B_{[2]}} - e^{-A_{[2]}}\,A_{[3]}\,B_{[1]}\,e^{-B_{[2]}}\,B_{[3]} + e^{-A_{[2]}}\,e^{-B_{[2]}}\,B_{[3]}$$

])

> **mrec:=map(expand,evalm(mm));**

mrec :=

$$\left[\left(e^{A_{[2]}}\right)^{4}\left(e^{V_{[2]}}\right)^{2} VV_{[1]} - 2\left(e^{A_{[2]}}\right)^{2}\left(e^{V_{[2]}}\right)^{2} VV_{[1]}\,A_{[1]}\,A_{[3]} \right.$$

$$+ \left(e^{V_{[2]}}\right)^2 VV_{[1]} A_{[1]}^{\,2} A_{[3]}^{\,2} + 2\left(e^{A_{[2]}}\right)^2 \left(e^{V_{[2]}}\right)^2 VV_{[1]} A_{[3]} RR_{[1]}$$

$$- 2\left(e^{V_{[2]}}\right)^2 VV_{[1]} A_{[3]}^{\,2} RR_{[1]} A_{[1]} + \left(e^{V_{[2]}}\right)^2 VV_{[1]} A_{[3]}^{\,2} RR_{[1]}^{\,2}$$

$$+ \left(e^{A_{[2]}}\right)^2 \left(e^{V_{[2]}}\right)^2 VV_{[2]} A_{[3]} - \left(e^{V_{[2]}}\right)^2 VV_{[2]} A_{[3]}^{\,2} A_{[1]}$$

$$+ \left(e^{V_{[2]}}\right)^2 VV_{[2]} A_{[3]}^{\,2} RR_{[1]} + \left(e^{V_{[2]}}\right)^2 A_{[3]}^{\,2} VV_{[3]} \left(e^{RR_{[2]}}\right)^2 \Bigg]$$

$$\Bigg[-2\left(e^{A_{[2]}}\right)^4 \left(e^{V_{[2]}}\right)^2 VV_{[1]} V_{[1]} + 4\left(e^{A_{[2]}}\right)^2 \left(e^{V_{[2]}}\right)^2 VV_{[1]} V_{[1]} A_{[1]} A_{[3]}$$

$$- 2\left(e^{V_{[2]}}\right)^2 VV_{[1]} V_{[1]} A_{[1]}^{\,2} A_{[3]}^{\,2} - 2\left(e^{A_{[2]}}\right)^2 \left(e^{V_{[2]}}\right)^2 VV_{[1]} A_{[1]}$$

$$+ 2\left(e^{V_{[2]}}\right)^2 VV_{[1]} A_{[1]}^{\,2} A_{[3]} - 4\left(e^{A_{[2]}}\right)^2 \left(e^{V_{[2]}}\right)^2 VV_{[1]} RR_{[1]} V_{[1]} A_{[3]}$$

$$+ 4\left(e^{V_{[2]}}\right)^2 VV_{[1]} RR_{[1]} V_{[1]} A_{[3]}^{\,2} A_{[1]} + 2\left(c^{A_{[2]}}\right)^2 \left(e^{V_{[2]}}\right)^2 VV_{[1]} RR_{[1]}$$

$$- 4\left(e^{V_{[2]}}\right)^2 VV_{[1]} RR_{[1]} A_{[1]} A_{[3]} - 2\left(e^{V_{[2]}}\right)^2 VV_{[1]} RR_{[1]}^{\,2} V_{[1]} A_{[3]}^{\,2}$$

$$+ 2\left(e^{V_{[2]}}\right)^2 VV_{[1]} RR_{[1]}^{\,2} A_{[3]} - 2\left(e^{A_{[2]}}\right)^2 \left(e^{V_{[2]}}\right)^2 VV_{[2]} V_{[1]} A_{[3]}$$

$$+ 2\left(e^{V_{[2]}}\right)^2 VV_{[2]} V_{[1]} A_{[3]}^{\,2} A_{[1]} + \left(e^{A_{[2]}}\right)^2 \left(e^{V_{[2]}}\right)^2 VV_{[2]}$$

$$- 2\left(e^{V_{[2]}}\right)^2 VV_{[2]} A_{[1]} A_{[3]} - 2\left(e^{V_{[2]}}\right)^2 VV_{[2]} RR_{[1]} V_{[1]} A_{[3]}^{\,2}$$

$$
+2\left(e^{V_{[2]}}\right)^2 VV_{[2]} A_{[3]} RR_{[1]} - 2\left(e^{V_{[2]}}\right)^2 A_{[3]}{}^2 VV_{[3]} V_{[1]} \left(e^{RR_{[2]}}\right)^2
$$

$$
+2\left(e^{V_{[2]}}\right)^2 A_{[3]} VV_{[3]} \left(e^{RR_{[2]}}\right)^2 \Bigg]
$$

$$
\Bigg[-2 VV_{[1]} RR_{[1]} V_{[1]} \left(e^{A_{[2]}}\right)^2 + VV_{[2]} V_{[1]}{}^2 A_{[3]} \left(e^{A_{[2]}}\right)^2
$$

$$
-2 VV_{[1]} V_{[1]}{}^2 A_{[1]} A_{[3]} \left(e^{A_{[2]}}\right)^2 + \left(e^{RR_{[2]}}\right)^2 VV_{[3]} V_{[1]}{}^2 A_{[3]}{}^2
$$

$$
+2 VV_{[1]} RR_{[1]} V_{[1]}{}^2 A_{[3]} \left(e^{A_{[2]}}\right)^2 - 2\left(e^{RR_{[2]}}\right)^2 VV_{[3]} V_{[1]} A_{[3]}
$$

$$
+2 VV_{[1]} V_{[1]} A_{[1]} \left(e^{A_{[2]}}\right)^2 + VV_{[1]} V_{[1]}{}^2 \left(e^{A_{[2]}}\right)^4 - VV_{[2]} V_{[1]} \left(e^{A_{[2]}}\right)^2
$$

$$
+\left(e^{RR_{[2]}}\right)^2 VV_{[3]} - VV_{[2]} V_{[1]}{}^2 A_{[3]}{}^2 A_{[1]} + 2 VV_{[2]} V_{[1]} A_{[1]} A_{[3]}
$$

$$
+ VV_{[2]} RR_{[1]} V_{[1]}{}^2 A_{[3]}{}^2 - 2 VV_{[2]} RR_{[1]} V_{[1]} A_{[3]}
$$

$$
+ VV_{[1]} V_{[1]}{}^2 A_{[1]}{}^2 A_{[3]}{}^2 - 2 VV_{[1]} V_{[1]} A_{[1]}{}^2 A_{[3]}
$$

$$
- 2 VV_{[1]} RR_{[1]} V_{[1]}{}^2 A_{[3]}{}^2 A_{[1]} + 4 VV_{[1]} RR_{[1]} V_{[1]} A_{[1]} A_{[3]}
$$

$$
+ VV_{[1]} RR_{[1]}{}^2 V_{[1]}{}^2 A_{[3]}{}^2 - 2 VV_{[1]} RR_{[1]}{}^2 V_{[1]} A_{[3]} + VV_{[2]} RR_{[1]}
$$

$$
- 2 VV_{[1]} RR_{[1]} A_{[1]} + VV_{[1]} A_{[1]}{}^2 + VV_{[1]} RR_{[1]}{}^2 - VV_{[2]} A_{[1]} \Bigg]
$$

```
> kirlv();
```

$$
\begin{bmatrix}
0 & -2\,x_{[1]} & -x_{[2]} \\
2\,x_{[1]} & 0 & -2\,x_{[3]} \\
x_{[2]} & 2\,x_{[3]} & 0
\end{bmatrix}
$$

> **duals();**

$$\text{xi-dagger}_{[1]} = \delta_{[1]}$$

$$\text{xi-dagger}_{[2]} = 2\,A_{[1]}\,\delta_{[1]} + \delta_{[2]}$$

$$\text{xi-dagger}_{[3]} = A_{[1]}^{\,2}\,\delta_{[1]} + A_{[1]}\,\delta_{[2]} + e^{2\,A_{[2]}}\,\delta_{[3]}$$

$$\text{xi-star}_{[1]} = e^{2\,A_{[2]}}\,\delta_{[1]} + A_{[3]}\,\delta_{[2]} + A_{[3]}^{\,2}\,\delta_{[3]}$$

$$\text{xi-star}_{[2]} = \delta_{[2]} + 2\,A_{[3]}\,\delta_{[3]}$$

$$\text{xi-star}_{[3]} = \delta_{[3]}$$

$$\text{xi-hat}_{[1]} = R_{[1]}$$

$$\text{xi-hat}_{[2]} = 2\,V_{[1]}\,R_{[1]} + R_{[2]}$$

$$\text{xi-hat}_{[3]} = V_{[1]}^{\,2}\,R_{[1]} + V_{[1]}\,R_{[2]} + e^{2\,V_{[2]}}\,R_{[3]}$$

> **indrep(1);**

$$\text{inducedrep}_{[1]} = R_{[1]}$$

$$\text{inducedrep}_{[2]} = 2\,V_{[1]}\,R_{[1]} + t$$

$$\text{inducedrep}_{[3]} = V_{[1]}^{\,2}\,R_{[1]} + V_{[1]}\,t$$

This is so(3), rotations in three dimensions.

> X:=matrix(3,3,[[0,a[3],-a[2]],[-a[3],0,a[1]],[a[2],-a[1],0]]);

$$X := \begin{bmatrix} 0 & a_{[3]} & -a_{[2]} \\ -a_{[3]} & 0 & a_{[1]} \\ a_{[2]} & -a_{[1]} & 0 \end{bmatrix}$$

> d:=3:n:=3:lie(X,d,n,1);

exp-adjoint

$$\left[\cos\!\left(A_{[3]}\right)\cos\!\left(A_{[2]}\right),\; \sin\!\left(A_{[3]}\right)\cos\!\left(A_{[2]}\right),\; -\sin\!\left(A_{[2]}\right) \right]$$
$$\left[-\cos\!\left(A_{[1]}\right)\sin\!\left(A_{[3]}\right) + \sin\!\left(A_{[1]}\right)\cos\!\left(A_{[3]}\right)\sin\!\left(A_{[2]}\right),\right.$$
$$\left. \cos\!\left(A_{[1]}\right)\cos\!\left(A_{[3]}\right) + \sin\!\left(A_{[1]}\right)\sin\!\left(A_{[2]}\right)\sin\!\left(A_{[3]}\right),\; \sin\!\left(A_{[1]}\right)\cos\!\left(A_{[2]}\right) \right]$$
$$\left[\sin\!\left(A_{[1]}\right)\sin\!\left(A_{[3]}\right) + \cos\!\left(A_{[1]}\right)\cos\!\left(A_{[3]}\right)\sin\!\left(A_{[2]}\right),\right.$$
$$\left. -\sin\!\left(A_{[1]}\right)\cos\!\left(A_{[3]}\right) + \cos\!\left(A_{[1]}\right)\sin\!\left(A_{[2]}\right)\sin\!\left(A_{[3]}\right),\; \cos\!\left(A_{[2]}\right)\cos\!\left(A_{[1]}\right) \right]$$

left-dual

$$\begin{bmatrix} 1 & 0 & 0 \\ \dfrac{\sin\!\left(A_{[2]}\right)\sin\!\left(A_{[1]}\right)}{\cos\!\left(A_{[2]}\right)} & \cos\!\left(A_{[1]}\right) & \dfrac{\sin\!\left(A_{[1]}\right)}{\cos\!\left(A_{[2]}\right)} \\ \dfrac{\sin\!\left(A_{[2]}\right)\cos\!\left(A_{[1]}\right)}{\cos\!\left(A_{[2]}\right)} & -\sin\!\left(A_{[1]}\right) & \dfrac{\cos\!\left(A_{[1]}\right)}{\cos\!\left(A_{[2]}\right)} \end{bmatrix}$$

right-dual

$$\begin{bmatrix} \dfrac{\cos\!\left(A_{[3]}\right)}{\cos\!\left(A_{[2]}\right)} & -\sin\!\left(A_{[3]}\right) & \dfrac{\cos\!\left(A_{[3]}\right)\sin\!\left(A_{[2]}\right)}{\cos\!\left(A_{[2]}\right)} \\ \dfrac{\sin\!\left(A_{[3]}\right)}{\cos\!\left(A_{[2]}\right)} & \cos\!\left(A_{[3]}\right) & \dfrac{\sin\!\left(A_{[2]}\right)\sin\!\left(A_{[3]}\right)}{\cos\!\left(A_{[2]}\right)} \\ 0 & 0 & 1 \end{bmatrix}$$

```
> mrec:=map(expand, evalm(mm));
```

$mrec :=$

$$\left[VV_{[1]} \cos\!\left(A_{[3]}\right) \cos\!\left(A_{[2]}\right) \cos\!\left(RR_{[2]}\right) \right.$$

$$- VV_{[1]} \sin\!\left(RR_{[2]}\right) \sin\!\left(RR_{[1]}\right) \cos\!\left(A_{[1]}\right) \sin\!\left(A_{[3]}\right)$$

$$+ VV_{[1]} \sin\!\left(RR_{[2]}\right) \sin\!\left(RR_{[1]}\right) \sin\!\left(A_{[1]}\right) \cos\!\left(A_{[3]}\right) \sin\!\left(A_{[2]}\right)$$

$$+ VV_{[1]} \sin\!\left(RR_{[2]}\right) \cos\!\left(RR_{[1]}\right) \sin\!\left(A_{[1]}\right) \sin\!\left(A_{[3]}\right)$$

$$+ VV_{[1]} \sin\!\left(RR_{[2]}\right) \cos\!\left(RR_{[1]}\right) \cos\!\left(A_{[1]}\right) \cos\!\left(A_{[3]}\right) \sin\!\left(A_{[2]}\right)$$

$$- VV_{[2]} \cos\!\left(RR_{[2]}\right) \cos\!\left(RR_{[1]}\right) \cos\!\left(A_{[1]}\right) \sin\!\left(A_{[3]}\right)$$

$$+ VV_{[2]} \cos\!\left(RR_{[2]}\right) \cos\!\left(RR_{[1]}\right) \sin\!\left(A_{[1]}\right) \cos\!\left(A_{[3]}\right) \sin\!\left(A_{[2]}\right)$$

$$- VV_{[2]} \cos\!\left(RR_{[2]}\right) \sin\!\left(RR_{[1]}\right) \sin\!\left(A_{[1]}\right) \sin\!\left(A_{[3]}\right)$$

$$- VV_{[2]} \cos\!\left(RR_{[2]}\right) \sin\!\left(RR_{[1]}\right) \cos\!\left(A_{[1]}\right) \cos\!\left(A_{[3]}\right) \sin\!\left(A_{[2]}\right)$$

$$- VV_{[3]} \sin\!\left(RR_{[1]}\right) \cos\!\left(A_{[1]}\right) \sin\!\left(A_{[3]}\right)$$

$$+ VV_{[3]} \sin\!\left(RR_{[1]}\right) \sin\!\left(A_{[1]}\right) \cos\!\left(A_{[3]}\right) \sin\!\left(A_{[2]}\right)$$

$$+ VV_{[3]} \cos\!\left(RR_{[1]}\right) \sin\!\left(A_{[1]}\right) \sin\!\left(A_{[3]}\right)$$

$$\left. + VV_{[3]} \cos\!\left(RR_{[1]}\right) \cos\!\left(A_{[1]}\right) \cos\!\left(A_{[3]}\right) \sin\!\left(A_{[2]}\right) \right]$$

$$\left[VV_{[1]} \cos\!\left(RR_{[2]}\right) \cos\!\left(V_{[1]}\right) \sin\!\left(A_{[3]}\right) \cos\!\left(A_{[2]}\right) \right.$$

$$+ VV_{[1]} \cos\!\left(RR_{[2]}\right) \sin\!\left(V_{[1]}\right) \sin\!\left(A_{[2]}\right)$$

$$+ VV_{[1]} \sin\!\left(RR_{[2]}\right) \sin\!\left(RR_{[1]}\right) \cos\!\left(V_{[1]}\right) \cos\!\left(A_{[1]}\right) \cos\!\left(A_{[3]}\right)$$

$$+ VV_{[1]} \sin\!\left(RR_{[2]}\right) \sin\!\left(RR_{[1]}\right) \cos\!\left(V_{[1]}\right) \sin\!\left(A_{[1]}\right) \sin\!\left(A_{[2]}\right) \sin\!\left(A_{[3]}\right)$$

$$- VV_{[1]} \sin\!\left(RR_{[2]}\right) \sin\!\left(RR_{[1]}\right) \sin\!\left(V_{[1]}\right) \sin\!\left(A_{[1]}\right) \cos\!\left(A_{[2]}\right)$$

$$- VV_{[1]} \sin\!\left(RR_{[2]}\right) \cos\!\left(RR_{[1]}\right) \cos\!\left(V_{[1]}\right) \sin\!\left(A_{[1]}\right) \cos\!\left(A_{[3]}\right)$$

$$+ VV_{[1]} \sin\!\left(RR_{[2]}\right) \cos\!\left(RR_{[1]}\right) \cos\!\left(V_{[1]}\right) \cos\!\left(A_{[1]}\right) \sin\!\left(A_{[2]}\right) \sin\!\left(A_{[3]}\right)$$

$$- VV_{[1]} \sin\!\left(RR_{[2]}\right) \cos\!\left(RR_{[1]}\right) \sin\!\left(V_{[1]}\right) \cos\!\left(A_{[2]}\right) \cos\!\left(A_{[1]}\right)$$

$$+ VV_{[2]} \cos\!\left(RR_{[2]}\right) \cos\!\left(RR_{[1]}\right) \cos\!\left(V_{[1]}\right) \cos\!\left(A_{[1]}\right) \cos\!\left(A_{[3]}\right)$$

$$+ VV_{[2]} \cos\!\left(RR_{[2]}\right) \cos\!\left(RR_{[1]}\right) \cos\!\left(V_{[1]}\right) \sin\!\left(A_{[1]}\right) \sin\!\left(A_{[2]}\right) \sin\!\left(A_{[3]}\right)$$

$$- VV_{[2]} \cos\!\left(RR_{[2]}\right) \cos\!\left(RR_{[1]}\right) \sin\!\left(V_{[1]}\right) \sin\!\left(A_{[1]}\right) \cos\!\left(A_{[2]}\right)$$

$$+ VV_{[2]} \cos\!\left(RR_{[2]}\right) \sin\!\left(RR_{[1]}\right) \cos\!\left(V_{[1]}\right) \sin\!\left(A_{[1]}\right) \cos\!\left(A_{[3]}\right)$$

$$- VV_{[2]} \cos\!\left(RR_{[2]}\right) \sin\!\left(RR_{[1]}\right) \cos\!\left(V_{[1]}\right) \cos\!\left(A_{[1]}\right) \sin\!\left(A_{[2]}\right) \sin\!\left(A_{[3]}\right)$$

$$+ VV_{[2]} \cos\!\left(RR_{[2]}\right) \sin\!\left(RR_{[1]}\right) \sin\!\left(V_{[1]}\right) \cos\!\left(A_{[2]}\right) \cos\!\left(A_{[1]}\right)$$

$$+ VV_{[3]} \sin\!\left(RR_{[1]}\right) \cos\!\left(V_{[1]}\right) \cos\!\left(A_{[1]}\right) \cos\!\left(A_{[3]}\right)$$

$$+ VV_{[3]} \sin\!\left(RR_{[1]}\right) \cos\!\left(V_{[1]}\right) \sin\!\left(A_{[1]}\right) \sin\!\left(A_{[2]}\right) \sin\!\left(A_{[3]}\right)$$

$$- VV_{[3]} \sin\!\left(RR_{[1]}\right) \sin\!\left(V_{[1]}\right) \sin\!\left(A_{[1]}\right) \cos\!\left(A_{[2]}\right)$$

$$- VV_{[3]} \cos\!\left(RR_{[1]}\right) \cos\!\left(V_{[1]}\right) \sin\!\left(A_{[1]}\right) \cos\!\left(A_{[3]}\right)$$

$$+ VV_{[3]} \cos\!\left(RR_{[1]}\right) \cos\!\left(V_{[1]}\right) \cos\!\left(A_{[1]}\right) \sin\!\left(A_{[2]}\right) \sin\!\left(A_{[3]}\right)$$

$$- VV_{[3]} \cos\!\left(RR_{[1]}\right) \sin\!\left(V_{[1]}\right) \cos\!\left(A_{[2]}\right) \cos\!\left(A_{[1]}\right)]$$

$$\left[-VV_{[3]} \sin\!\left(RR_{[1]}\right) \sin\!\left(V_{[2]}\right) \sin\!\left(A_{[1]}\right) \cos\!\left(A_{[3]}\right) \sin\!\left(A_{[2]}\right)\right.$$

$$- VV_{[3]} \cos\!\left(RR_{[1]}\right) \sin\!\left(V_{[2]}\right) \cos\!\left(A_{[1]}\right) \cos\!\left(A_{[3]}\right) \sin\!\left(A_{[2]}\right)$$

$$- VV_{[2]} \cos\!\left(RR_{[2]}\right) \cos\!\left(RR_{[1]}\right) \sin\!\left(V_{[2]}\right) \sin\!\left(A_{[1]}\right) \cos\!\left(A_{[3]}\right) \sin\!\left(A_{[2]}\right)$$

$$+ VV_{[2]} \cos\!\left(RR_{[2]}\right) \cos\!\left(RR_{[1]}\right) \sin\!\left(V_{[2]}\right) \cos\!\left(A_{[1]}\right) \sin\!\left(A_{[3]}\right)$$

$$- VV_{[1]} \sin\!\left(RR_{[2]}\right) \cos\!\left(RR_{[1]}\right) \sin\!\left(V_{[2]}\right) \cos\!\left(A_{[1]}\right) \cos\!\left(A_{[3]}\right) \sin\!\left(A_{[2]}\right)$$

$$+ VV_{[2]} \cos\!\left(RR_{[2]}\right) \sin\!\left(RR_{[1]}\right) \sin\!\left(V_{[2]}\right) \sin\!\left(A_{[1]}\right) \sin\!\left(A_{[3]}\right)$$

$$+ VV_{[2]} \cos\!\left(RR_{[2]}\right) \sin\!\left(RR_{[1]}\right) \sin\!\left(V_{[2]}\right) \cos\!\left(A_{[1]}\right) \cos\!\left(A_{[3]}\right) \sin\!\left(A_{[2]}\right)$$

$$+ VV_{[3]} \sin\!\left(RR_{[1]}\right) \sin\!\left(V_{[2]}\right) \cos\!\left(A_{[1]}\right) \sin\!\left(A_{[3]}\right)$$

$$- VV_{[1]} \cos\!\left(RR_{[2]}\right) \sin\!\left(V_{[2]}\right) \cos\!\left(A_{[3]}\right) \cos\!\left(A_{[2]}\right)$$

$$+ VV_{[1]} \sin\!\left(RR_{[2]}\right) \sin\!\left(RR_{[1]}\right) \sin\!\left(V_{[2]}\right) \cos\!\left(A_{[1]}\right) \sin\!\left(A_{[3]}\right)$$

$$- VV_{[3]} \cos\!\left(RR_{[1]}\right) \sin\!\left(V_{[2]}\right) \sin\!\left(A_{[1]}\right) \sin\!\left(A_{[3]}\right)$$

$$- VV_{[1]} \sin\!\left(RR_{[2]}\right) \cos\!\left(RR_{[1]}\right) \sin\!\left(V_{[2]}\right) \sin\!\left(A_{[1]}\right) \sin\!\left(A_{[3]}\right)$$

$$- VV_{[1]} \sin\!\left(RR_{[2]}\right) \sin\!\left(RR_{[1]}\right) \sin\!\left(V_{[2]}\right) \sin\!\left(A_{[1]}\right) \cos\!\left(A_{[3]}\right) \sin\!\left(A_{[2]}\right)$$

$$+ VV_{[3]} \cos\!\left(RR_{[1]}\right) \cos\!\left(V_{[1]}\right) \cos\!\left(V_{[2]}\right) \cos\!\left(A_{[2]}\right) \cos\!\left(A_{[1]}\right)$$

$$+ VV_{[2]} \cos\!\left(RR_{[2]}\right) \sin\!\left(RR_{[1]}\right) \sin\!\left(V_{[1]}\right) \cos\!\left(V_{[2]}\right) \sin\!\left(A_{[1]}\right) \cos\!\left(A_{[3]}\right)$$

$$- VV_{[2]} \cos\!\left(RR_{[2]}\right) \sin\!\left(RR_{[1]}\right) \cos\!\left(V_{[1]}\right) \cos\!\left(V_{[2]}\right) \cos\!\left(A_{[2]}\right) \cos\!\left(A_{[1]}\right)$$

$$+ VV_{[1]} \sin\!\left(RR_{[2]}\right) \sin\!\left(RR_{[1]}\right) \sin\!\left(V_{[1]}\right) \cos\!\left(V_{[2]}\right) \cos\!\left(A_{[1]}\right) \cos\!\left(A_{[3]}\right)$$

$$+ VV_{[1]} \sin\left(RR_{[2]}\right) \sin\left(RR_{[1]}\right) \sin\left(V_{[1]}\right) \cos\left(V_{[2]}\right) \sin\left(A_{[1]}\right) \sin\left(A_{[2]}\right) \sin\left(A_{[3]}\right)$$

$$+ VV_{[1]} \sin\left(RR_{[2]}\right) \sin\left(RR_{[1]}\right) \cos\left(V_{[1]}\right) \cos\left(V_{[2]}\right) \sin\left(A_{[1]}\right) \cos\left(A_{[2]}\right)$$

$$+ VV_{[2]} \cos\left(RR_{[2]}\right) \cos\left(RR_{[1]}\right) \cos\left(V_{[1]}\right) \cos\left(V_{[2]}\right) \sin\left(A_{[1]}\right) \cos\left(A_{[2]}\right) +$$

$$VV_{[1]} \sin\left(RR_{[2]}\right) \cos\left(RR_{[1]}\right) \sin\left(V_{[1]}\right) \cos\left(V_{[2]}\right) \cos\left(A_{[1]}\right) \sin\left(A_{[2]}\right) \sin\left(A_{[3]}\right)$$

$$+ VV_{[2]} \cos\left(RR_{[2]}\right) \cos\left(RR_{[1]}\right) \sin\left(V_{[1]}\right) \cos\left(V_{[2]}\right) \cos\left(A_{[1]}\right) \cos\left(A_{[3]}\right)$$

$$+ VV_{[1]} \sin\left(RR_{[2]}\right) \cos\left(RR_{[1]}\right) \cos\left(V_{[1]}\right) \cos\left(V_{[2]}\right) \cos\left(A_{[2]}\right) \cos\left(A_{[1]}\right) -$$

$$VV_{[2]} \cos\left(RR_{[2]}\right) \sin\left(RR_{[1]}\right) \sin\left(V_{[1]}\right) \cos\left(V_{[2]}\right) \cos\left(A_{[1]}\right) \sin\left(A_{[2]}\right) \sin\left(A_{[3]}\right)$$

$$+ VV_{[3]} \sin\left(RR_{[1]}\right) \sin\left(V_{[1]}\right) \cos\left(V_{[2]}\right) \sin\left(A_{[1]}\right) \sin\left(A_{[2]}\right) \sin\left(A_{[3]}\right)$$

$$- VV_{[3]} \cos\left(RR_{[1]}\right) \sin\left(V_{[1]}\right) \cos\left(V_{[2]}\right) \sin\left(A_{[1]}\right) \cos\left(A_{[3]}\right)$$

$$+ VV_{[3]} \cos\left(RR_{[1]}\right) \sin\left(V_{[1]}\right) \cos\left(V_{[2]}\right) \cos\left(A_{[1]}\right) \sin\left(A_{[2]}\right) \sin\left(A_{[3]}\right)$$

$$+ VV_{[3]} \sin\left(RR_{[1]}\right) \sin\left(V_{[1]}\right) \cos\left(V_{[2]}\right) \cos\left(A_{[1]}\right) \cos\left(A_{[3]}\right)$$

$$+ VV_{[3]} \sin\left(RR_{[1]}\right) \cos\left(V_{[1]}\right) \cos\left(V_{[2]}\right) \sin\left(A_{[1]}\right) \cos\left(A_{[2]}\right) +$$

$$VV_{[2]} \cos\left(RR_{[2]}\right) \cos\left(RR_{[1]}\right) \sin\left(V_{[1]}\right) \cos\left(V_{[2]}\right) \sin\left(A_{[1]}\right) \sin\left(A_{[2]}\right) \sin\left(A_{[3]}\right)$$

$$+ VV_{[1]} \cos\left(RR_{[2]}\right) \sin\left(V_{[1]}\right) \cos\left(V_{[2]}\right) \sin\left(A_{[3]}\right) \cos\left(A_{[2]}\right)$$

$$- VV_{[1]} \cos\left(RR_{[2]}\right) \cos\left(V_{[1]}\right) \cos\left(V_{[2]}\right) \sin\left(A_{[2]}\right)$$

$$- VV_{[1]} \sin\left(RR_{[2]}\right) \cos\left(RR_{[1]}\right) \sin\left(V_{[1]}\right) \cos\left(V_{[2]}\right) \sin\left(A_{[1]}\right) \cos\left(A_{[3]}\right) \Big]$$

> **kirlv();**

$$\begin{bmatrix} 0 & -x_{[3]} & x_{[2]} \\ x_{[3]} & 0 & -x_{[1]} \\ -x_{[2]} & x_{[1]} & 0 \end{bmatrix}$$

> duals();

$$xi\text{-}dagger_{[1]} = \delta_{[1]}$$

$$xi\text{-}dagger_{[2]} = \frac{\sin\left(A_{[2]}\right)\sin\left(A_{[1]}\right)\delta_{[1]} + \cos\left(A_{[1]}\right)\delta_{[2]}\cos\left(A_{[2]}\right) + \sin\left(A_{[1]}\right)\delta_{[3]}}{\cos\left(A_{[2]}\right)}$$

$$xi\text{-}dagger_{[3]} =$$
$$-\frac{-\sin\left(A_{[2]}\right)\cos\left(A_{[1]}\right)\delta_{[1]} + \sin\left(A_{[1]}\right)\delta_{[2]}\cos\left(A_{[2]}\right) - \cos\left(A_{[1]}\right)\delta_{[3]}}{\cos\left(A_{[2]}\right)}$$

$$xi\text{-}star_{[1]} = -\frac{-\cos\left(A_{[3]}\right)\delta_{[1]} + \sin\left(A_{[3]}\right)\delta_{[2]}\cos\left(A_{[2]}\right) - \cos\left(A_{[3]}\right)\sin\left(A_{[2]}\right)\delta_{[3]}}{\cos\left(A_{[2]}\right)}$$

$$xi\text{-}star_{[2]} = \frac{\sin\left(A_{[3]}\right)\delta_{[1]} + \cos\left(A_{[3]}\right)\delta_{[2]}\cos\left(A_{[2]}\right) + \sin\left(A_{[2]}\right)\sin\left(A_{[3]}\right)\delta_{[3]}}{\cos\left(A_{[2]}\right)}$$

$$xi\text{-}star_{[3]} = \delta_{[3]}$$

$$xi\text{-}hat_{[1]} = R_{[1]}$$

$$xi\text{-}hat_{[2]} = \frac{\sin\left(V_{[2]}\right)\sin\left(V_{[1]}\right)R_{[1]} + \cos\left(V_{[1]}\right)R_{[2]}\cos\left(V_{[2]}\right) + \sin\left(V_{[1]}\right)R_{[3]}}{\cos\left(V_{[2]}\right)}$$

$$xi\text{-}hat_{[3]} =$$
$$-\frac{-\sin\left(V_{[2]}\right)\cos\left(V_{[1]}\right)R_{[1]} + \sin\left(V_{[1]}\right)R_{[2]}\cos\left(V_{[2]}\right) - \cos\left(V_{[1]}\right)R_{[3]}}{\cos\left(V_{[2]}\right)}$$

This is the complex Euclidean group in two dimensions.

> X:=matrix(3,3,[[0,a[2],I*a[3]-I*a[1]],[-a[2],0,a[1]+a[3]],[0,0,0]]);

$$X := \begin{bmatrix} 0 & a_{[2]} & I\,a_{[3]} - I\,a_{[1]} \\ -a_{[2]} & 0 & a_{[1]} + a_{[3]} \\ 0 & 0 & 0 \end{bmatrix}$$

> d:=3:n:=3:lie(X,d,n,0);

group element

$$\begin{bmatrix} \dfrac{1}{2}\,e^{I\,A_{[2]}} + \dfrac{1}{2}\,\dfrac{1}{e^{I\,A_{[2]}}} & -\dfrac{1}{2}\,I\,e^{I\,A_{[2]}} + \dfrac{1}{2}\,\dfrac{I}{e^{I\,A_{[2]}}} & \dfrac{I\,A_{[3]}}{e^{I\,A_{[2]}}} - I\,A_{[1]} \\[3ex] \dfrac{1}{2}\,I\,e^{I\,A_{[2]}} - \dfrac{1}{2}\,\dfrac{I}{e^{I\,A_{[2]}}} & \dfrac{1}{2}\,e^{I\,A_{[2]}} + \dfrac{1}{2}\,\dfrac{1}{e^{I\,A_{[2]}}} & \dfrac{A_{[3]}}{e^{I\,A_{[2]}}} + A_{[1]} \\[3ex] 0 & 0 & 1 \end{bmatrix}$$

exp-adjoint

$$\begin{bmatrix} e^{I\,A_{[2]}} & -I\,A_{[1]} & 0 \\[2ex] 0 & 1 & 0 \\[2ex] 0 & I\,A_{[3]}\,e^{-I\,A_{[2]}} & e^{-I\,A_{[2]}} \end{bmatrix}$$

left-dual

$$\begin{bmatrix} 1 & 0 & 0 \\ IA_{[1]} & 1 & 0 \\ & & IA_{[2]} \\ 0 & 0 & e \end{bmatrix}$$

right-dual

$$\begin{bmatrix} IA_{[2]} & & \\ e & 0 & 0 \\ 0 & 1 & IA_{[3]} \\ 0 & 0 & 1 \end{bmatrix}$$

> **matrec();**

$$\begin{bmatrix} IA_{[2]} \\ e & VV_{[1]} \end{bmatrix}$$

$$\begin{bmatrix} -I\,VV_{[1]}\,V_{[1]}\,e^{IA_{[2]}} - I\,VV_{[1]}\,A_{[1]} + I\,VV_{[1]}\,RR_{[1]} + VV_{[2]} \\[2ex] + \dfrac{IA_{[3]}\,VV_{[3]}\,e^{I\,RR_{[2]}}}{e^{IA_{[2]}}} \end{bmatrix}$$

$$\begin{bmatrix} \dfrac{VV_{[3]}\,e^{I\,RR_{[2]}}}{e^{I\,V_{[2]}}\,e^{IA_{[2]}}} \end{bmatrix}$$

> kirlv();

$$\begin{bmatrix} 0 & -I\,x_{[1]} & 0 \\ I\,x_{[1]} & 0 & -I\,x_{[3]} \\ 0 & I\,x_{[3]} & 0 \end{bmatrix}$$

> duals();

$$xi\text{-}dagger_{[1]} = \delta_{[1]}$$

$$xi\text{-}dagger_{[2]} = I\,A_{[1]}\,\delta_{[1]} + \delta_{[2]}$$

$$xi\text{-}dagger_{[3]} = e^{I\,A_{[2]}}\,\delta_{[3]}$$

$$xi\text{-}star_{[1]} = e^{I\,A_{[2]}}\,\delta_{[1]}$$

$$xi\text{-}star_{[2]} = \delta_{[2]} + I\,A_{[3]}\,\delta_{[3]}$$

$$xi\text{-}star_{[3]} = \delta_{[3]}$$

$$xi\text{-}hat_{[1]} = R_{[1]}$$

$$xi\text{-}hat_{[2]} = I\,V_{[1]}\,R_{[1]} + R_{[2]}$$

$$xi\text{-}hat_{[3]} = e^{I\,V_{[2]}}\,R_{[3]}$$

Here is the Heisenberg algebra. The Heisenberg group is the same as N3.

> X:=matrix(3,3,[0,a[3],a[2],0,0,a[1],0,0,0]);

$$X := \begin{bmatrix} 0 & a_{[3]} & a_{[2]} \\ 0 & 0 & a_{[1]} \\ 0 & 0 & 0 \end{bmatrix}$$

> d:=3:n:=3:lie(X,d,n,0);

group element

$$\begin{bmatrix} 1 & A_{[3]} & A_{[2]} \\ 0 & 1 & A_{[1]} \\ 0 & 0 & 1 \end{bmatrix}$$

exp-adjoint

$$\begin{bmatrix} 1 & 0 & 0 \\ A_{[3]} & 1 & -A_{[1]} \\ 0 & 0 & 1 \end{bmatrix}$$

left-dual

$$\begin{bmatrix} 1 & 0 & 0 \\ 0 & 1 & 0 \\ 0 & A_{[1]} & 1 \end{bmatrix}$$

right-dual

$$\begin{bmatrix} 1 & A_{[3]} & 0 \\ 0 & 1 & 0 \\ 0 & 0 & 1 \end{bmatrix}$$

> **grp();**

group – entries

table([

$1 = B_{[1]} + A_{[1]}$

$2 = B_{[2]} + A_{[3]} B_{[1]} + A_{[2]}$

$3 = B_{[3]} + A_{[3]}$

])

> **matrec();**

$$\begin{bmatrix} VV_{[1]} + A_{[3]} VV_{[2]} \\ VV_{[2]} \\ -VV_{[2]} V_{[1]} - VV_{[2]} A_{[1]} + VV_{[2]} RR_{[1]} + VV_{[3]} \end{bmatrix}$$

> **kirlv();**

$$\begin{bmatrix} 0 & 0 & -x_{[2]} \\ 0 & 0 & 0 \\ x_{[2]} & 0 & 0 \end{bmatrix}$$

> duals();

$$xi\text{-}dagger_{[1]} = \delta_{[1]}$$

$$xi\text{-}dagger_{[2]} = \delta_{[2]}$$

$$xi\text{-}dagger_{[3]} = A_{[1]}\,\delta_{[2]} + \delta_{[3]}$$

$$xi\text{-}star_{[1]} = \delta_{[1]} + A_{[3]}\,\delta_{[2]}$$

$$xi\text{-}star_{[2]} = \delta_{[2]}$$

$$xi\text{-}star_{[3]} = \delta_{[3]}$$

$$xi\text{-}hat_{[1]} = R_{[1]}$$

$$xi\text{-}hat_{[2]} = R_{[2]}$$

$$xi\text{-}hat_{[3]} = V_{[1]}\,R_{[2]} + R_{[3]}$$

> indrep(1);

$$inducedrep_{[1]} = R_{[1]}$$

$$inducedrep_{[2]} = t$$

$$inducedrep_{[3]} = V_{[1]}\,t$$

This is the seven-dimensional Heisenberg group, H7.

```
> X:=matrix(5,5,[0,a[7],a[6],a[5],a[4],0,0,0,0,a[3],0,0,0
> ,0,a[2],0,0,0,0,a[1],0,0,0,0,0]);
```

$$X := \begin{bmatrix} 0 & a_{[7]} & a_{[6]} & a_{[5]} & a_{[4]} \\ 0 & 0 & 0 & 0 & a_{[3]} \\ 0 & 0 & 0 & 0 & a_{[2]} \\ 0 & 0 & 0 & 0 & a_{[1]} \\ 0 & 0 & 0 & 0 & 0 \end{bmatrix}$$

```
> d:=7:n:=5:lie(X,d,n,0);
```

group element

$$\begin{bmatrix} 1 & A_{[7]} & A_{[6]} & A_{[5]} & A_{[4]} \\ 0 & 1 & 0 & 0 & A_{[3]} \\ 0 & 0 & 1 & 0 & A_{[2]} \\ 0 & 0 & 0 & 1 & A_{[1]} \\ 0 & 0 & 0 & 0 & 1 \end{bmatrix}$$

exp-adjoint

$$\begin{bmatrix} 1 & 0 & 0 & 0 & 0 & 0 & 0 \\ 0 & 1 & 0 & 0 & 0 & 0 & 0 \\ 0 & 0 & 1 & 0 & 0 & 0 & 0 \\ A_{[5]} & A_{[6]} & A_{[7]} & 1 & -A_{[1]} & -A_{[2]} & -A_{[3]} \\ 0 & 0 & 0 & 0 & 1 & 0 & 0 \\ 0 & 0 & 0 & 0 & 0 & 1 & 0 \\ 0 & 0 & 0 & 0 & 0 & 0 & 1 \end{bmatrix}$$

left-dual

$$\begin{bmatrix} 1 & 0 & 0 & 0 & 0 & 0 & 0 \\ 0 & 1 & 0 & 0 & 0 & 0 & 0 \\ 0 & 0 & 1 & 0 & 0 & 0 & 0 \\ 0 & 0 & 0 & 1 & 0 & 0 & 0 \\ 0 & 0 & 0 & A_{[1]} & 1 & 0 & 0 \\ 0 & 0 & 0 & A_{[2]} & 0 & 1 & 0 \\ 0 & 0 & 0 & A_{[3]} & 0 & 0 & 1 \end{bmatrix}$$

right-dual

$$\begin{bmatrix} 1 & 0 & 0 & A_{[5]} & 0 & 0 & 0 \\ 0 & 1 & 0 & A_{[6]} & 0 & 0 & 0 \\ 0 & 0 & 1 & A_{[7]} & 0 & 0 & 0 \\ 0 & 0 & 0 & 1 & 0 & 0 & 0 \\ 0 & 0 & 0 & 0 & 1 & 0 & 0 \\ 0 & 0 & 0 & 0 & 0 & 1 & 0 \\ 0 & 0 & 0 & 0 & 0 & 0 & 1 \end{bmatrix}$$

```
> grp();
```

group – entries

table([

$1 = B_{[1]} + A_{[1]}$

$2 = B_{[2]} + A_{[2]}$

$3 = B_{[3]} + A_{[3]}$

$4 = B_{[4]} + A_{[7]} B_{[3]} + A_{[6]} B_{[2]} + A_{[5]} B_{[1]} + A_{[4]}$

$5 = B_{[5]} + A_{[5]}$

$6 = B_{[6]} + A_{[6]}$

$7 = B_{[7]} + A_{[7]}$

])

```
> matrec();
```

$$
\begin{bmatrix}
VV_{[1]} + A_{[5]}\, VV_{[4]} \\
VV_{[2]} + A_{[6]}\, VV_{[4]} \\
VV_{[3]} + A_{[7]}\, VV_{[4]} \\
VV_{[4]} \\
-VV_{[4]}\, V_{[1]} - VV_{[4]}\, A_{[1]} + VV_{[4]}\, RR_{[1]} + VV_{[5]} \\
-VV_{[4]}\, V_{[2]} - VV_{[4]}\, A_{[2]} + VV_{[4]}\, RR_{[2]} + VV_{[6]} \\
-VV_{[4]}\, V_{[3]} - VV_{[4]}\, A_{[3]} + VV_{[4]}\, RR_{[3]} + VV_{[7]}
\end{bmatrix}
$$

```
> kirlv();
```

$$
\begin{bmatrix}
0 & 0 & 0 & 0 & -x_{[4]} & 0 & 0 \\
0 & 0 & 0 & 0 & 0 & -x_{[4]} & 0 \\
0 & 0 & 0 & 0 & 0 & 0 & -x_{[4]} \\
0 & 0 & 0 & 0 & 0 & 0 & 0 \\
x_{[4]} & 0 & 0 & 0 & 0 & 0 & 0 \\
0 & x_{[4]} & 0 & 0 & 0 & 0 & 0 \\
0 & 0 & x_{[4]} & 0 & 0 & 0 & 0
\end{bmatrix}
$$

```
> duals();
```

$$
xi\text{-}dagger_{[1]} = \delta_{[1]}
$$

$$
xi\text{-}dagger_{[2]} = \delta_{[2]}
$$

$$
xi\text{-}dagger_{[3]} = \delta_{[3]}
$$

$$
xi\text{-}dagger_{[4]} = \delta_{[4]}
$$

$$
xi\text{-}dagger_{[5]} = A_{[1]}\, \delta_{[4]} + \delta_{[5]}
$$

$$
xi\text{-}dagger_{[6]} = A_{[2]}\, \delta_{[4]} + \delta_{[6]}
$$

$$
xi\text{-}dagger_{[7]} = A_{[3]}\, \delta_{[4]} + \delta_{[7]}
$$

$$xi\text{-}star_{[1]} = \delta_{[1]} + A_{[5]}\,\delta_{[4]}$$

$$xi\text{-}star_{[2]} = \delta_{[2]} + A_{[6]}\,\delta_{[4]}$$

$$xi\text{-}star_{[3]} = \delta_{[3]} + A_{[7]}\,\delta_{[4]}$$

$$xi\text{-}star_{[4]} = \delta_{[4]}$$

$$\acute{x}i\text{-}star_{[5]} = \delta_{[5]}$$

$$xi\text{-}star_{[6]} = \delta_{[6]}$$

$$xi\text{-}star_{[7]} = \delta_{[7]}$$

$$xi\text{-}hat_{[1]} = R_{[1]}$$

$$xi\text{-}hat_{[2]} = R_{[2]}$$

$$xi\text{-}hat_{[3]} = R_{[3]}$$

$$xi\text{-}hat_{[4]} = R_{[4]}$$

$$xi\text{-}hat_{[5]} = V_{[1]}\,R_{[4]} + R_{[5]}$$

$$xi\text{-}hat_{[6]} = V_{[2]}\,R_{[4]} + R_{[6]}$$

$$xi\text{-}hat_{[7]} = V_{[3]}\,R_{[4]} + R_{[7]}$$

```
> indrep(3);
```

$$inducedrep_{[1]} = R_{[1]}$$

$$inducedrep_{[2]} = R_{[2]}$$

$$inducedrep_{[3]} = R_{[3]}$$

$$inducedrep_{[4]} = t$$

$$inducedrep_{[5]} = V_{[1]} t$$

$$inducedrep_{[6]} = V_{[2]} t$$

$$inducedrep_{[7]} = V_{[3]} t$$

This is the group of 2x2 upper-triangular matrices. The Lie algebra is isomorphic to the finite-difference algebra.

```
> X:=matrix(2,2,[[a[3],a[2]],[0,a[1]]]);
```

$$X := \begin{bmatrix} a_{[3]} & a_{[2]} \\ 0 & a_{[1]} \end{bmatrix}$$

```
> d:=3:n:=2:lie(X,d,n,0);
```

group element

$$\begin{bmatrix} e^{A_{[3]}} & A_{[2]} \\ 0 & e^{A_{[1]}} \end{bmatrix}$$

exp-adjoint

$$\begin{bmatrix} 1 & 0 & 0 \\ A_{[2]}\, e^{-A_{[1]}} & e^{-A_{[1]}+A_{[3]}} & -A_{[2]}\, e^{-A_{[1]}} \\ 0 & 0 & 1 \end{bmatrix}$$

left-dual

$$\begin{bmatrix} 1 & 0 & 0 \\ 0 & e^{A_{[1]}} & 0 \\ 0 & A_{[2]} & 1 \end{bmatrix}$$

right-dual

$$\begin{bmatrix} 1 & A_{[2]} & 0 \\ 0 & e^{A_{[3]}} & 0 \\ 0 & 0 & 1 \end{bmatrix}$$

```
> grp();
```

$$group - entries$$

table([

$$1 = e^{A_{[1]}} e^{B_{[1]}}$$

$$2 = e^{A_{[3]}} B_{[2]} + A_{[2]} e^{B_{[1]}}$$

$$3 = e^{A_{[3]}} e^{B_{[3]}}$$

])

```
> matrec();
```

$$
\begin{bmatrix}
VV_{[1]} + \dfrac{A_{[2]} \, VV_{[2]} \, e^{RR_{[1]}}}{e^{A_{[1]}}} \\[2em]
\dfrac{VV_{[2]} \, e^{A_{[3]}} \, e^{RR_{[1]}}}{e^{V_{[1]}} \, e^{A_{[1]}}} \\[2em]
-\dfrac{VV_{[2]} \, V_{[2]} \, e^{RR_{[1]}} \, e^{A_{[3]}}}{e^{V_{[1]}} \, e^{A_{[1]}}} - \dfrac{A_{[2]} \, VV_{[2]} \, e^{RR_{[1]}}}{e^{A_{[1]}}} + VV_{[2]} \, RR_{[2]} + VV_{[3]}
\end{bmatrix}
$$

```
> kirlv();
```

$$
\begin{bmatrix}
0 & -x_{[2]} & 0 \\
x_{[2]} & 0 & -x_{[2]} \\
0 & x_{[2]} & 0
\end{bmatrix}
$$

```
> duals();
```

$$xi\text{-}dagger_{[1]} = \delta_{[1]}$$

$$xi\text{-}dagger_{[2]} = e^{A_{[1]}} \delta_{[2]}$$

$$xi\text{-}dagger_{[3]} = A_{[2]} \delta_{[2]} + \delta_{[3]}$$

$$xi\text{-}star_{[1]} = \delta_{[1]} + A_{[2]} \hat{o}_{[2]}$$

$$xi\text{-}star_{[2]} = e^{A_{[3]}} \delta_{[2]}$$

$$xi\text{-}star_{[3]} = \delta_{[3]}$$

$$xi\text{-}hat_{[1]} = R_{[1]}$$

$$xi\text{-}hat_{[2]} = e^{V_{[1]}} R_{[2]}$$

$$xi\text{-}hat_{[3]} = V_{[2]} R_{[2]} + R_{[3]}$$

```
> indrep(2);
```

$$inducedrep_{[1]} = R_{[1]}$$

$$inducedrep_{[2]} = e^{V_{[1]}} R_{[2]}$$

$$inducedrep_{[3]} = V_{[2]} R_{[2]} + t$$

This is the oscillator group. The Lie algebra consists of the Heisenberg algebra plus the number operator.

> **X:=matrix(3,3,[[0,a[4],a[2]],[0,a[3],a[1]],[0,0,0]]);**

$$X := \begin{bmatrix} 0 & a_{[4]} & a_{[2]} \\ 0 & a_{[3]} & a_{[1]} \\ 0 & 0 & 0 \end{bmatrix}$$

> **d:=4:n:=3:lie(X,d,n,0);**

group element

$$\begin{bmatrix} 1 & A_{[4]} & A_{[2]} \\ 0 & e^{A_{[3]}} & A_{[1]} \\ 0 & 0 & 1 \end{bmatrix}$$

exp-adjoint

$$\begin{bmatrix} e^{A_{[3]}} & 0 & -A_{[1]} & 0 \\ A_{[4]} & 1 & -A_{[1]} e^{-A_{[3]}} & A_{[4]} & -A_{[1]} e^{-A_{[3]}} \\ 0 & 0 & 1 & 0 \\ 0 & 0 & e^{-A_{[3]}} A_{[4]} & e^{-A_{[3]}} \end{bmatrix}$$

left-dual

$$\begin{bmatrix} 1 & 0 & 0 & 0 \\ 0 & 1 & 0 & 0 \\ A_{[1]} & 0 & 1 & 0 \\ 0 & A_{[1]} & 0 & e^{A_{[3]}} \end{bmatrix}$$

right-dual

$$\begin{bmatrix} e^{A_{[3]}} & A_{[4]} & 0 & 0 \\ 0 & 1 & 0 & 0 \\ 0 & 0 & 1 & A_{[4]} \\ 0 & 0 & 0 & 1 \end{bmatrix}$$

> **grp();**

group – entries

table([

$$3 = e^{A_{[3]}} e^{B_{[3]}}$$

$$4 = B_{[4]} + A_{[4]} e^{B_{[3]}}$$

$$1 = e^{A_{[3]}} B_{[1]} + A_{[1]}$$

$$2 = B_{[2]} + A_{[4]} B_{[1]} + A_{[2]}$$

])

> **matrec();**

$$\begin{bmatrix} e^{A_{[3]}} VV_{[1]} + A_{[4]} VV_{[2]} \end{bmatrix}$$

$$\begin{bmatrix} VV_{[2]} \end{bmatrix}$$

$$\begin{bmatrix} -VV_{[1]} V_{[1]} e^{A_{[3]}} - VV_{[1]} A_{[1]} + VV_{[1]} RR_{[1]} - VV_{[2]} V_{[1]} A_{[4]} \end{bmatrix}$$

$$-\frac{VV_{[2]}\,A_{[1]}\,A_{[4]}}{A_{e}^{[3]}}+\frac{VV_{[2]}\,A_{[4]}\,RR_{[1]}}{A_{e}^{[3]}}+VV_{[3]}+\frac{A_{[4]}\,VV_{[4]}\,e^{RR_{[3]}}}{A_{e}^{[3]}}$$

$$\left[-\frac{VV_{[2]}\,V_{[1]}}{V_{e}^{[3]}}-\frac{VV_{[2]}\,A_{[1]}}{V_{e}^{[3]}\,A_{e}^{[3]}}+\frac{VV_{[2]}\,RR_{[1]}}{V_{e}^{[3]}\,A_{e}^{[3]}}+\frac{VV_{[4]}\,e^{RR_{[3]}}}{V_{e}^{[3]}\,A_{e}^{[3]}}\right]$$

> kirlv();

$$\begin{bmatrix} 0 & 0 & -x_{[1]} & -x_{[2]} \\ 0 & 0 & 0 & 0 \\ x_{[1]} & 0 & 0 & -x_{[4]} \\ x_{[2]} & 0 & x_{[4]} & 0 \end{bmatrix}$$

> duals();

$$xi\text{-}dagger_{[1]} = \delta_{[1]}$$

$$xi\text{-}dagger_{[2]} = \delta_{[2]}$$

$$xi\text{-}dagger_{[3]} = A_{[1]}\,\delta_{[1]} + \delta_{[3]}$$

$$xi\text{-}dagger_{[4]} = A_{[1]}\,\delta_{[2]} + e^{A_{[3]}}\,\delta_{[4]}$$

$$xi\text{-}star_{[1]} = e^{A_{[3]}}\,\delta_{[1]} + A_{[4]}\,\delta_{[2]}$$

$$xi\text{-}star_{[2]} = \delta_{[2]}$$

$$xi\text{-}star_{[3]} = \delta_{[3]} + A_{[4]}\,\delta_{[4]}$$

$$xi\text{-}star_{[4]} = \delta_{[4]}$$

$$xi\text{-}hat_{[1]} = R_{[1]}$$

$$xi\text{-}hat_{[2]} = R_{[2]}$$

$$xi\text{-}hat_{[3]} = V_{[1]} R_{[1]} + R_{[3]}$$

$$xi\text{-}hat_{[4]} = V_{[1]} R_{[2]} + e^{V_{[3]}} R_{[4]}$$

> **indrep(1);**

$$inducedrep_{[1]} = R_{[1]}$$

$$inducedrep_{[2]} = t$$

$$inducedrep_{[3]} = V_{[1]} R_{[1]}$$

$$inducedrep_{[4]} = V_{[1]} t$$

This is a four-dimensional nilpotent group. The Lie algebra is isomorphic to that generated by D and multiplication by x^2/2.

> X:=matrix(4,4,[[0,a[4],0,a[3]],[0,0,a[4],a[2]],[0,0,0,a[1]],[0,0,0,0]]);

$$X := \begin{bmatrix} 0 & a_{[4]} & 0 & a_{[3]} \\ 0 & 0 & a_{[4]} & a_{[2]} \\ 0 & 0 & 0 & a_{[1]} \\ 0 & 0 & 0 & 0 \end{bmatrix}$$

> d:=4:n:=4:lie(X,d,n,0);

group element

$$\begin{bmatrix} 1 & A_{[4]} & \frac{1}{2}A_{[4]}^{2} & A_{[3]} \\ 0 & 1 & A_{[4]} & A_{[2]} \\ 0 & 0 & 1 & A_{[1]} \\ 0 & 0 & 0 & 1 \end{bmatrix}$$

exp-adjoint

$$\begin{bmatrix} 1 & 0 & 0 & 0 \\ A_{[4]} & 1 & 0 & -A_{[1]} \\ \frac{1}{2}A_{[4]}^{2} & A_{[4]} & 1 & -A_{[2]} \\ 0 & 0 & 0 & 1 \end{bmatrix}$$

left-dual

$$\begin{bmatrix} 1 & 0 & 0 & 0 \\ 0 & 1 & 0 & 0 \\ 0 & 0 & 1 & 0 \\ 0 & A_{[1]} & A_{[2]} & 1 \end{bmatrix}$$

right-dual

$$\begin{bmatrix} 1 & A_{[4]} & \frac{1}{2}A_{[4]}^{2} & 0 \\ 0 & 1 & A_{[4]} & 0 \\ 0 & 0 & 1 & 0 \\ 0 & 0 & 0 & 1 \end{bmatrix}$$

> grp();

group – entries

table([

$1 = B_{[1]} + A_{[1]}$

$2 = B_{[2]} + A_{[4]} B_{[1]} + A_{[2]}$

$3 = B_{[3]} + A_{[4]} B_{[2]} + \frac{1}{2}A_{[4]}^{2} B_{[1]} + A_{[3]}$

$4 = B_{[4]} + A_{[4]}$

])

> matrec();

$$\left[VV_{[1]} + A_{[4]} \overset{\downarrow}{V}V_{[2]} + \frac{1}{2}A_{[4]}^{2} VV_{[3]} \right]$$
$$\left[VV_{[2]} + A_{[4]} VV_{[3]} \right]$$
$$\left[VV_{[3]} \right]$$
$$\left[-VV_{[2]} V_{[1]} - VV_{[2]} A_{[1]} + VV_{[2]} RR_{[1]} - VV_{[3]} V_{[1]} A_{[4]} - VV_{[3]} V_{[2]} \right.$$
$$\left. - VV_{[3]} A_{[2]} + VV_{[3]} RR_{[2]} + VV_{[4]} \right]$$

> kirlv();

$$\begin{bmatrix} 0 & 0 & 0 & -x_{[2]} \\ 0 & 0 & 0 & -x_{[3]} \\ 0 & 0 & 0 & 0 \\ x_{[2]} & x_{[3]} & 0 & 0 \end{bmatrix}$$

> duals();

$$xi\text{-}dagger_{[1]} = \delta_{[1]}$$

$$xi\text{-}dagger_{[2]} = \delta_{[2]}$$

$$xi\text{-}dagger_{[3]} = \delta_{[3]}$$

$$xi\text{-}dagger_{[4]} = A_{[1]}\,\delta_{[2]} + A_{[2]}\,\delta_{[3]} + \delta_{[4]}$$

$$xi\text{-}star_{[1]} = \delta_{[1]} + A_{[4]}\,\delta_{[2]} + \frac{1}{2}A_{[4]}^{2}\,\delta_{[3]}$$

$$xi\text{-}star_{[2]} = \delta_{[2]} + A_{[4]}\,\delta_{[3]}$$

$$xi\text{-}star_{[3]} = \delta_{[3]}$$

$$xi\text{-}star_{[4]} = \delta_{[4]}$$

$$xi\text{-}hat_{[1]} = R_{[1]}$$

$$xi\text{-}hat_{[2]} = R_{[2]}$$

$$xi\text{-}hat_{[3]} = R_{[3]}$$

$$xi\text{-}hat_{[4]} = V_{[1]}\,R_{[2]} + V_{[2]}\,R_{[3]} + R_{[4]}$$

> indrep(2);

$$inducedrep_{[1]} = R_{[1]}$$

$$inducedrep_{[2]} = R_{[2]}$$

$$inducedrep_{[3]} = t$$

$$inducedrep_{[4]} = V_{[1]}\,R_{[2]} + V_{[2]}\,t$$

This is N5, the algebra of strictly upper-triangular matrices.

```
> X:=matrix(5,5,[0,a[10],a[9],a[7],a[4],0,0,a[8],a[6],a[
> ],0,0,0,a[5],a[2],0,0,0,0,a[1],0,0,0,0,0]);
```

$$X := \begin{bmatrix} 0 & a_{[10]} & a_{[9]} & a_{[7]} & a_{[4]} \\ 0 & 0 & a_{[8]} & a_{[6]} & a_{[3]} \\ 0 & 0 & 0 & a_{[5]} & a_{[2]} \\ 0 & 0 & 0 & 0 & a_{[1]} \\ 0 & 0 & 0 & 0 & 0 \end{bmatrix}$$

```
> d:=10:n:=5:lie(X,d,n,0);
```

group element

$$\begin{bmatrix} 1 & A_{[10]} & A_{[9]} & A_{[7]} & A_{[4]} \\ 0 & 1 & A_{[8]} & A_{[6]} & A_{[3]} \\ 0 & 0 & 1 & A_{[5]} & A_{[2]} \\ 0 & 0 & 0 & 1 & A_{[1]} \\ 0 & 0 & 0 & 0 & 1 \end{bmatrix}$$

exp-adjoint

$$[1,0,0,0,0,0,0,0,0,0]$$
$$\left[A_{[5]},1,0,0,-A_{[1]},0,0,0,0,0\right]$$
$$\left[A_{[6]},A_{[8]},1,0,-A_{[1]}A_{[8]},-A_{[1]},0,-A_{[2]}+A_{[1]}A_{[5]},0,0\right]$$
$$\left[A_{[7]},A_{[9]},A_{[10]},1,-A_{[1]}A_{[9]},-A_{[1]}A_{[10]},-A_{[1]},\right.$$
$$-\left(A_{[2]}-A_{[1]}A_{[5]}\right)A_{[10]},-A_{[2]}+A_{[1]}A_{[5]},$$
$$\left.-A_{[3]}+A_{[1]}A_{[6]}+A_{[8]}A_{[2]}-A_{[8]}A_{[1]}A_{[5]}\right]$$
$$[0,0,0,0,1,0,0,0,0,0]$$
$$\left[0,0,0,0,A_{[8]},1,0,-A_{[5]},0,0\right]$$

$$\left[0,0,0,0,A_{[9]},A_{[10]},1,-A_{[5]}A_{[10]},-A_{[5]},-A_{[6]}+A_{[5]}A_{[8]}\right]$$
$$[0,0,0,0,0,0,0,1,0,0]$$
$$\left[0,0,0,0,0,0,0,A_{[10]},1,-A_{[8]}\right]$$
$$[0,0,0,0,0,0,0,0,0,1]$$

left-dual

$$\begin{bmatrix}
1 & 0 & 0 & 0 & 0 & 0 & 0 & 0 & 0 & 0 \\
0 & 1 & 0 & 0 & 0 & 0 & 0 & 0 & 0 & 0 \\
0 & 0 & 1 & 0 & 0 & 0 & 0 & 0 & 0 & 0 \\
0 & 0 & 0 & 1 & 0 & 0 & 0 & 0 & 0 & 0 \\
0 & A_{[1]} & 0 & 0 & 1 & 0 & 0 & 0 & 0 & 0 \\
0 & 0 & A_{[1]} & 0 & 0 & 1 & 0 & 0 & 0 & 0 \\
0 & 0 & 0 & A_{[1]} & 0 & 0 & 1 & 0 & 0 & 0 \\
0 & 0 & A_{[2]} & 0 & 0 & A_{[5]} & 0 & 1 & 0 & 0 \\
0 & 0 & 0 & A_{[2]} & 0 & 0 & A_{[5]} & 0 & 1 & 0 \\
0 & 0 & 0 & A_{[3]} & 0 & 0 & A_{[6]} & 0 & A_{[8]} & 1
\end{bmatrix}$$

right-dual

$$\begin{bmatrix}
1 & A_{[5]} & A_{[6]} & A_{[7]} & 0 & 0 & 0 & 0 & 0 & 0 \\
0 & 1 & A_{[8]} & A_{[9]} & 0 & 0 & 0 & 0 & 0 & 0 \\
0 & 0 & 1 & A_{[10]} & 0 & 0 & 0 & 0 & 0 & 0 \\
0 & 0 & 0 & 1 & 0 & 0 & 0 & 0 & 0 & 0 \\
0 & 0 & 0 & 0 & 1 & A_{[8]} & A_{[9]} & 0 & 0 & 0 \\
0 & 0 & 0 & 0 & 0 & 1 & A_{[10]} & 0 & 0 & 0 \\
0 & 0 & 0 & 0 & 0 & 0 & 1 & 0 & 0 & 0 \\
0 & 0 & 0 & 0 & 0 & 0 & 0 & 1 & A_{[10]} & 0 \\
0 & 0 & 0 & 0 & 0 & 0 & 0 & 0 & 1 & 0 \\
0 & 0 & 0 & 0 & 0 & 0 & 0 & 0 & 0 & 1
\end{bmatrix}$$

```
> grp();
```

group – entries

```
table([
```

$$4 = B_{[4]} + A_{[10]} B_{[3]} + A_{[9]} B_{[2]} + A_{[7]} B_{[1]} + A_{[4]}$$

$$5 = B_{[5]} + A_{[5]}$$

$$6 = B_{[6]} + A_{[8]} B_{[5]} + A_{[6]}$$

$$7 = B_{[7]} + A_{[10]} B_{[6]} + A_{[9]} B_{[5]} + A_{[7]}$$

$$8 = B_{[8]} + A_{[8]}$$

$$9 = B_{[9]} + A_{[10]} B_{[8]} + A_{[9]}$$

$$10 = B_{[10]} + A_{[10]}$$

$$1 = B_{[1]} + A_{[1]}$$

$$2 = B_{[2]} + A_{[5]} B_{[1]} + A_{[2]}$$

$$3 = B_{[3]} + A_{[8]} B_{[2]} + A_{[6]} B_{[1]} + A_{[3]}$$

```
])
```

```
> matrec();
```

$$\left[VV_{[1]} + A_{[5]} VV_{[2]} + A_{[6]} VV_{[3]} + A_{[7]} VV_{[4]} \right]$$

$$\left[VV_{[2]} + A_{[8]} VV_{[3]} + A_{[9]} VV_{[4]} \right]$$

$$\left[VV_{[3]} + A_{[10]} VV_{[4]} \right]$$

$$\left[VV_{[4]} \right]$$

$$\left[-VV_{[2]} V_{[1]} - VV_{[2]} A_{[1]} + VV_{[2]} RR_{[1]} - VV_{[3]} V_{[1]} A_{[8]} - VV_{[3]} A_{[1]} A_{[8]} \right.$$
$$+ VV_{[3]} A_{[8]} RR_{[1]} - VV_{[4]} V_{[1]} A_{[9]} - VV_{[4]} A_{[1]} A_{[9]} + VV_{[4]} A_{[9]} RR_{[1]}$$
$$\left. + VV_{[5]} + A_{[8]} VV_{[6]} + A_{[9]} VV_{[7]} \right]$$

$$\left[-VV_{[3]} V_{[1]} - VV_{[3]} A_{[1]} + VV_{[3]} RR_{[1]} - VV_{[4]} V_{[1]} A_{[10]} \right.$$
$$\left. - VV_{[4]} A_{[1]} A_{[10]} + VV_{[4]} A_{[10]} RR_{[1]} + VV_{[6]} + A_{[10]} VV_{[7]} \right]$$

$$\left[-VV_{[4]} V_{[1]} - VV_{[4]} A_{[1]} + VV_{[4]} RR_{[1]} + VV_{[7]} \right]$$

$$\left[-VV_{[3]} V_{[2]} + VV_{[3]} V_{[5]} V_{[1]} + VV_{[3]} V_{[5]} A_{[1]} - VV_{[3]} A_{[2]} \right.$$
$$+ VV_{[3]} A_{[1]} A_{[5]} - VV_{[3]} RR_{[1]} V_{[5]} - VV_{[3]} RR_{[1]} A_{[5]} + VV_{[3]} RR_{[2]}$$
$$- VV_{[4]} A_{[10]} V_{[2]} + VV_{[4]} A_{[10]} V_{[5]} V_{[1]} + VV_{[4]} V_{[5]} A_{[1]} A_{[10]}$$
$$\left. - VV_{[4]} A_{[10]} A_{[2]} + VV_{[4]} A_{[10]} A_{[1]} A_{[5]} - VV_{[4]} RR_{[1]} V_{[5]} A_{[10]} \right.$$

$$- VV_{[4]} RR_{[1]} A_{[5]} A_{[10]} + VV_{[4]} A_{[10]} RR_{[2]} - VV_{[6]} V_{[5]} - VV_{[6]} A_{[5]}$$
$$+ VV_{[6]} RR_{[5]} - VV_{[7]} V_{[5]} A_{[10]} - VV_{[7]} A_{[5]} A_{[10]} + VV_{[7]} A_{[10]} RR_{[5]}$$
$$+ VV_{[8]} + A_{[10]} VV_{[9]}]$$

$$[-VV_{[4]} V_{[2]} + VV_{[4]} V_{[5]} V_{[1]} + VV_{[4]} V_{[5]} A_{[1]} - VV_{[4]} A_{[2]}$$
$$+ VV_{[4]} A_{[1]} A_{[5]} - VV_{[4]} RR_{[1]} V_{[5]} - VV_{[4]} RR_{[1]} A_{[5]} + VV_{[4]} RR_{[2]}$$
$$- VV_{[7]} V_{[5]} - VV_{[7]} A_{[5]} + VV_{[7]} RR_{[5]} + VV_{[9]}]$$

$$[VV_{[10]} + VV_{[4]} A_{[1]} V_{[6]} + VV_{[4]} V_{[8]} A_{[2]} + VV_{[4]} A_{[1]} A_{[6]}$$
$$+ VV_{[4]} A_{[8]} A_{[2]} - VV_{[4]} RR_{[1]} V_{[6]} - VV_{[4]} RR_{[1]} A_{[6]} - VV_{[4]} RR_{[2]} V_{[8]}$$
$$- VV_{[4]} RR_{[2]} A_{[8]} + VV_{[7]} V_{[8]} V_{[5]} + VV_{[7]} V_{[8]} A_{[5]} + VV_{[7]} A_{[5]} A_{[8]}$$
$$- VV_{[7]} RR_{[5]} V_{[8]} - VV_{[7]} RR_{[5]} A_{[8]} - VV_{[4]} V_{[3]} - VV_{[4]} A_{[3]}$$
$$+ VV_{[4]} RR_{[3]} + VV_{[4]} V_{[6]} V_{[1]} - VV_{[4]} V_{[8]} V_{[5]} V_{[1]}$$
$$- VV_{[4]} A_{[1]} V_{[8]} V_{[5]} - VV_{[4]} V_{[8]} A_{[1]} A_{[5]} - VV_{[4]} A_{[8]} A_{[1]} A_{[5]}$$
$$+ VV_{[4]} RR_{[1]} V_{[8]} V_{[5]} + VV_{[4]} RR_{[1]} V_{[8]} A_{[5]} + VV_{[4]} RR_{[1]} A_{[5]} A_{[8]}$$
$$- VV_{[7]} V_{[6]} - VV_{[7]} A_{[6]} + VV_{[7]} RR_{[6]} - VV_{[9]} V_{[8]} - VV_{[9]} A_{[8]}$$
$$+ VV_{[9]} RR_{[8]} + VV_{[4]} V_{[8]} V_{[2]}]$$

```
> kirlv();
```

$$\begin{bmatrix}
0 & 0 & 0 & 0 & -x_{[2]} & -x_{[3]} & -x_{[4]} & 0 & 0 & 0 \\
0 & 0 & 0 & 0 & 0 & 0 & 0 & -x_{[3]} & -x_{[4]} & 0 \\
0 & 0 & 0 & 0 & 0 & 0 & 0 & 0 & 0 & -x_{[4]} \\
0 & 0 & 0 & 0 & 0 & 0 & 0 & 0 & 0 & 0 \\
x_{[2]} & 0 & 0 & 0 & 0 & 0 & 0 & -x_{[6]} & -x_{[7]} & 0 \\
x_{[3]} & 0 & 0 & 0 & 0 & 0 & 0 & 0 & 0 & -x_{[7]} \\
x_{[4]} & 0 & 0 & 0 & 0 & 0 & 0 & 0 & 0 & 0 \\
0 & x_{[3]} & 0 & 0 & x_{[6]} & 0 & 0 & 0 & 0 & -x_{[9]} \\
0 & x_{[4]} & 0 & 0 & x_{[7]} & 0 & 0 & 0 & 0 & 0 \\
0 & 0 & x_{[4]} & 0 & 0 & x_{[7]} & 0 & x_{[9]} & 0 & 0
\end{bmatrix}$$

```
> duals();
```

$$xi\text{-}dagger_{[1]} = \delta_{[1]}$$

$$xi\text{-}dagger_{[2]} = \delta_{[2]}$$

$$xi\text{-}dagger_{[3]} = \delta_{[3]}$$

$$xi\text{-}dagger_{[4]} = \delta_{[4]}$$

$$xi\text{-}dagger_{[5]} = A_{[1]}\,\delta_{[2]} + \delta_{[5]}$$

$$xi\text{-}dagger_{[6]} = A_{[1]}\,\delta_{[3]} + \delta_{[6]}$$

$$xi\text{-}dagger_{[7]} = A_{[1]}\,\delta_{[4]} + \delta_{[7]}$$

$$xi\text{-}dagger_{[8]} = A_{[2]}\,\delta_{[3]} + A_{[5]}\,\delta_{[6]} + \delta_{[8]}$$

$$xi\text{-}dagger_{[9]} = A_{[2]}\,\delta_{[4]} + A_{[5]}\,\delta_{[7]} + \delta_{[9]}$$

$$xi\text{-}dagger_{[10]} = A_{[3]}\,\delta_{[4]} + A_{[6]}\,\delta_{[7]} + A_{[8]}\,\delta_{[9]} + \delta_{[10]}$$

$$xi\text{-}star_{[1]} = \delta_{[1]} + A_{[5]}\,\delta_{[2]} + A_{[6]}\,\delta_{[3]} + A_{[7]}\,\delta_{[4]}$$

$$xi\text{-}star_{[2]} = \delta_{[2]} + A_{[8]}\,\delta_{[3]} + A_{[9]}\,\delta_{[4]}$$

$$xi\text{-}star_{[3]} = \delta_{[3]} + A_{[10]}\,\delta_{[4]}$$

$$xi\text{-}star_{[4]} = \delta_{[4]}$$

$$xi\text{-}star_{[5]} = \delta_{[5]} + A_{[8]}\,\delta_{[6]} + A_{[9]}\,\delta_{[7]}$$

$$xi\text{-}star_{[6]} = \delta_{[6]} + A_{[10]}\,\delta_{[7]}$$

$$xi\text{-}star_{[7]} = \delta_{[7]}$$

$$xi\text{-}star_{[8]} = \delta_{[8]} + A_{[10]}\,\delta_{[9]}$$

$$xi\text{-}star_{[9]} = \delta_{[9]}$$

$$xi\text{-}star_{[10]} = \delta_{[10]}$$

$$xi\text{-}hat_{[1]} = R_{[1]}$$

$$xi\text{-}hat_{[2]} = R_{[2]}$$

$$xi\text{-}hat_{[3]} = R_{[3]}$$

$$xi\text{-}hat_{[4]} = R_{[4]}$$

$$xi\text{-}hat_{[5]} = V_{[1]} R_{[2]} + R_{[5]}$$

$$xi\text{-}hat_{[6]} = V_{[1]} R_{[3]} + R_{[6]}$$

$$xi\text{-}hat_{[7]} = V_{[1]} R_{[4]} + R_{[7]}$$

$$xi\text{-}hat_{[8]} = V_{[2]} R_{[3]} + V_{[5]} R_{[6]} + R_{[8]}$$

$$xi\text{-}hat_{[9]} = V_{[2]} R_{[4]} + V_{[5]} R_{[7]} + R_{[9]}$$

$$xi\text{-}hat_{[10]} = V_{[3]} R_{[4]} + V_{[6]} R_{[7]} + V_{[8]} R_{[9]} + R_{[10]}$$

```
> indrep(3);
```

$$inducedrep_{[1]} = R_{[1]}$$

$$inducedrep_{[2]} = R_{[2]}$$

$$inducedrep_{[3]} = R_{[3]}$$

$$inducedrep_{[4]} = t$$

$$inducedrep_{[5]} = V_{[1]} R_{[2]}$$

$$inducedrep_{[6]} = V_{[1]} R_{[3]}$$

$$inducedrep_{[7]} = V_{[1]} t$$

$$inducedrep_{[8]} = V_{[2]} R_{[3]}$$

$$inducedrep_{[9]} = V_{[2]} t$$

$$inducedrep_{[10]} = V_{[3]} t$$

```
>
```

This is the affine group in one dimension.

```
> X:=matrix(2,2,[a[2],a[1],0,0]);
```

$$X := \begin{bmatrix} a_{[2]} & a_{[1]} \\ 0 & 0 \end{bmatrix}$$

```
> d:=2:n:=2:lie(X,d,n,0);
```

group element

$$\begin{bmatrix} e^{A_{[2]}} & A_{[1]} \\ 0 & 1 \end{bmatrix}$$

exp-adjoint

$$\begin{bmatrix} e^{A_{[2]}} & -A_{[1]} \\ 0 & 1 \end{bmatrix}$$

left-dual

$$\begin{bmatrix} 1 & 0 \\ A_{[1]} & 1 \end{bmatrix}$$

right-dual

$$\begin{bmatrix} e^{A_{[2]}} & 0 \\ 0 & 1 \end{bmatrix}$$

```
> grp();
```

group − entries

table([

$$2 = e^{A_{[2]}} B_{[2]}$$

$$1 = e^{A_{[2]}} B_{[1]} + A_{[1]}$$

])

```
> matrec();
```

$$\begin{bmatrix} e^{A_{[2]}} \, VV_{[1]} \\ -VV_{[1]} \, V_{[1]} \, e^{A_{[2]}} - VV_{[1]} \, A_{[1]} + VV_{[1]} \, RR_{[1]} + VV_{[2]} \end{bmatrix}$$

```
> kirlv();
```

$$\begin{bmatrix} 0 & -x_{[1]} \\ x_{[1]} & 0 \end{bmatrix}$$

```
> duals();
```

$$xi\text{-}dagger_{[1]} = \delta_{[1]}$$

$$xi\text{-}dagger_{[2]} = A_{[1]} \, \delta_{[1]} + \delta_{[2]}$$

$$xi\text{-}star_{[1]} = e^{A_{[2]}} \, \delta_{[1]}$$

$$xi\text{-}star_{[2]} = \delta_{[2]}$$

$$xi\text{-}hat_{[1]} = R_{[1]}$$

$$xi\text{-}hat_{[2]} = V_{[1]} \, R_{[1]} + R_{[2]}$$

```
> indrep(1);
```

$$inducedrep_{[1]} = R_{[1]}$$

$$inducedrep_{[2]} = V_{[1]} \, R_{[1]} + t$$

This is the Euclidean group in three dimensions.

```
> X:=matrix(4,4,[0,a[6],-a[5],a[3],-a[6],0,a[4],a[2],a[5]
> ,-a[4],0,a[1],0,0,0,0]);
```

$$X := \begin{bmatrix} 0 & a_{[6]} & -a_{[5]} & a_{[3]} \\ -a_{[6]} & 0 & a_{[4]} & a_{[2]} \\ a_{[5]} & -a_{[4]} & 0 & a_{[1]} \\ 0 & 0 & 0 & 0 \end{bmatrix}$$

```
> d:=6:n:=4:lie(X,d,n,1);# See below for the group
> element.
```

exp-adjoint

$$\Big[\cos(A_{[4]})\cos(A_{[5]}),\ \cos(A_{[4]})\sin(A_{[5]})\sin(A_{[6]}) - \sin(A_{[4]})\cos(A_{[6]}),$$
$$\cos(A_{[4]})\sin(A_{[5]})\cos(A_{[6]}) + \sin(A_{[4]})\sin(A_{[6]}),\ A_{[3]}\cos(A_{[4]})\sin(A_{[6]})$$
$$+A_{[2]}\cos(A_{[6]})\cos(A_{[5]}) - \sin(A_{[5]})\cos(A_{[6]})A_{[3]}\sin(A_{[4]}),$$
$$-A_{[3]}\cos(A_{[4]})\cos(A_{[6]}) + A_{[2]}\sin(A_{[6]})\cos(A_{[5]})$$
$$- \sin(A_{[5]})\sin(A_{[6]})A_{[3]}\sin(A_{[4]}),\ -A_{[2]}\sin(A_{[5]}) - A_{[3]}\sin(A_{[4]})\cos(A_{[5]})$$

$$\Big[\sin(A_{[4]})\cos(A_{[5]}),\ \sin(A_{[4]})\sin(A_{[5]})\sin(A_{[6]}) + \cos(A_{[4]})\cos(A_{[6]}),$$
$$\sin(A_{[4]})\sin(A_{[5]})\cos(A_{[6]}) - \cos(A_{[4]})\sin(A_{[6]}),\ A_{[3]}\sin(A_{[4]})\sin(A_{[6]})$$
$$-A_{[1]}\cos(A_{[6]})\cos(A_{[5]}) + \sin(A_{[5]})\cos(A_{[6]})A_{[3]}\cos(A_{[4]}),$$
$$-A_{[3]}\sin(A_{[4]})\cos(A_{[6]}) - A_{[1]}\sin(A_{[6]})\cos(A_{[5]})$$
$$+ \sin(A_{[5]})\sin(A_{[6]})A_{[3]}\cos(A_{[4]}),\ A_{[1]}\sin(A_{[5]}) + A_{[3]}\cos(A_{[4]})\cos(A_{[5]})$$
$$\Big]$$

$$\Big[-\sin(A_{[5]}),\ \cos(A_{[5]})\sin(A_{[6]}),\ \cos(A_{[5]})\cos(A_{[6]}),\ -\sin(A_{[6]})A_{[1]}\cos(A_{[4]})$$
$$- \sin(A_{[6]})A_{[2]}\sin(A_{[4]}) + \sin(A_{[5]})\cos(A_{[6]})A_{[1]}\sin(A_{[4]})$$
$$- \sin(A_{[5]})\cos(A_{[6]})A_{[2]}\cos(A_{[4]}),\ \cos(A_{[6]})A_{[1]}\cos(A_{[4]})$$
$$+ \cos(A_{[6]})A_{[2]}\sin(A_{[4]}) + \sin(A_{[5]})\sin(A_{[6]})A_{[1]}\sin(A_{[4]})$$
$$- \sin(A_{[5]})\sin(A_{[6]})A_{[2]}\cos(A_{[4]}),$$

$$\cos\!\left(A_{[5]}\right) A_{[1]} \sin\!\left(A_{[4]}\right) - \cos\!\left(A_{[5]}\right) A_{[2]} \cos\!\left(A_{[4]}\right)\Big]$$

$$\left[0,0,0,\cos\!\left(A_{[5]}\right)\cos\!\left(A_{[6]}\right),\cos\!\left(A_{[5]}\right)\sin\!\left(A_{[6]}\right),-\sin\!\left(A_{[5]}\right)\right]$$

$$\left[0,0,0,\sin\!\left(A_{[4]}\right)\sin\!\left(A_{[5]}\right)\cos\!\left(A_{[6]}\right)-\cos\!\left(A_{[4]}\right)\sin\!\left(A_{[6]}\right),\right.$$

$$\left.\sin\!\left(A_{[4]}\right)\sin\!\left(A_{[5]}\right)\sin\!\left(A_{[6]}\right)+\cos\!\left(A_{[4]}\right)\cos\!\left(A_{[6]}\right),\sin\!\left(A_{[4]}\right)\cos\!\left(A_{[5]}\right)\right]$$

$$\left[0,0,0,\cos\!\left(A_{[4]}\right)\sin\!\left(A_{[5]}\right)\cos\!\left(A_{[6]}\right)+\sin\!\left(A_{[4]}\right)\sin\!\left(A_{[6]}\right),\right.$$

$$\left.\cos\!\left(A_{[4]}\right)\sin\!\left(A_{[5]}\right)\sin\!\left(A_{[6]}\right)-\sin\!\left(A_{[4]}\right)\cos\!\left(A_{[6]}\right),\cos\!\left(A_{[4]}\right)\cos\!\left(A_{[5]}\right)\right]$$

left-dual

$$
\begin{bmatrix}
1 & 0 & 0 & 0 & 0 & 0 \\
0 & 1 & 0 & 0 & 0 & 0 \\
0 & 0 & 1 & 0 & 0 & 0 \\
-A_{[2]} & A_{[1]} & 0 & 1 & 0 & 0 \\
A_{[3]} & 0 & -A_{[1]} & \dfrac{\sin\!\left(A_{[4]}\right)\sin\!\left(A_{[5]}\right)}{\cos\!\left(A_{[5]}\right)} & \cos\!\left(A_{[4]}\right) & \dfrac{\sin\!\left(A_{[4]}\right)}{\cos\!\left(A_{[5]}\right)} \\
0 & -A_{[3]} & A_{[2]} & \dfrac{\cos\!\left(A_{[4]}\right)\sin\!\left(A_{[5]}\right)}{\cos\!\left(A_{[5]}\right)} & -\sin\!\left(A_{[4]}\right) & \dfrac{\cos\!\left(A_{[4]}\right)}{\cos\!\left(A_{[5]}\right)}
\end{bmatrix}
$$

right-dual

$$\left[\cos\!\left(A_{[4]}\right)\cos\!\left(A_{[5]}\right),\sin\!\left(A_{[4]}\right)\cos\!\left(A_{[5]}\right),-\sin\!\left(A_{[5]}\right),0,0,0\right]$$

$$\left[\cos\!\left(A_{[4]}\right)\sin\!\left(A_{[5]}\right)\sin\!\left(A_{[6]}\right)-\sin\!\left(A_{[4]}\right)\cos\!\left(A_{[6]}\right),\right.$$

$$\left.\sin\!\left(A_{[4]}\right)\sin\!\left(A_{[5]}\right)\sin\!\left(A_{[6]}\right)+\cos\!\left(A_{[4]}\right)\cos\!\left(A_{[6]}\right),\cos\!\left(A_{[5]}\right)\sin\!\left(A_{[6]}\right),0,0,0\right]$$

$$\left[\cos\!\left(A_{[4]}\right)\sin\!\left(A_{[5]}\right)\cos\!\left(A_{[6]}\right)+\sin\!\left(A_{[4]}\right)\sin\!\left(A_{[6]}\right),\right.$$

$$\left.\sin\!\left(A_{[4]}\right)\sin\!\left(A_{[5]}\right)\cos\!\left(A_{[6]}\right)-\cos\!\left(A_{[4]}\right)\sin\!\left(A_{[6]}\right),\cos\!\left(A_{[5]}\right)\cos\!\left(A_{[6]}\right),0,0,0\right]$$

$$\left[0,0,0,\dfrac{\cos\!\left(A_{[6]}\right)}{\cos\!\left(A_{[5]}\right)},-\sin\!\left(A_{[6]}\right),\dfrac{\sin\!\left(A_{[5]}\right)\cos\!\left(A_{[6]}\right)}{\cos\!\left(A_{[5]}\right)}\right]$$

$$\left[0,0,0,\frac{\sin\left(A_{[6]}\right)}{\cos\left(A_{[5]}\right)},\cos\left(A_{[6]}\right),\frac{\sin\left(A_{[5]}\right)\sin\left(A_{[6]}\right)}{\cos\left(A_{[5]}\right)}\right]$$
$$[0,0,0,0,0,1]$$

The group element is converted to trigonometric form, as it is used in exponential form in the calculations.

```
> print(convert(hh,trig));
```

$$\left[\cos\left(A_{[5]}\right)\cos\left(A_{[6]}\right),\cos\left(A_{[5]}\right)\sin\left(A_{[6]}\right),-\sin\left(A_{[5]}\right),A_{[3]}\right]$$

$$\left[-\frac{1}{4}\cos(\%6)-\frac{1}{4}\cos(\%5)+\frac{1}{4}\cos(\%4)-\frac{1}{2}\sin(\%3)-\frac{1}{2}\sin(\%2)+\frac{1}{4}\cos(\%1),\right.$$

$$-\frac{1}{4}\sin(\%6)-\frac{1}{4}\sin(\%5)+\frac{1}{4}\sin(\%4)+\frac{1}{2}\cos(\%3)+\frac{1}{2}\cos(\%2)+\frac{1}{4}\sin(\%1),$$

$$\left.\sin\left(A_{[4]}\right)\cos\left(A_{[5]}\right),A_{[2]}\right]$$

$$\left[-\frac{1}{4}\sin(\%6)+\frac{1}{4}\sin(\%5)+\frac{1}{4}\sin(\%4)+\frac{1}{2}\cos(\%3)-\frac{1}{2}\cos(\%2)-\frac{1}{4}\sin(\%1),\right.$$

$$\frac{1}{4}\cos(\%6)-\frac{1}{4}\cos(\%5)-\frac{1}{4}\cos(\%4)+\frac{1}{2}\sin(\%3)-\frac{1}{2}\sin(\%2)+\frac{1}{4}\cos(\%1),$$

$$\left.\cos\left(A_{[4]}\right)\cos\left(A_{[5]}\right),A_{[1]}\right]$$
$$[0,0,0,1]$$

$$\%1 := A_{[4]}-A_{[5]}+A_{[6]}$$

$$\%2 := A_{[4]}+A_{[6]}$$

$$\%3 := -A_{[4]}+A_{[6]}$$

$$\%4 := -A_{[4]}+A_{[5]}+A_{[6]}$$

$$\%5 := A_{[4]}+A_{[5]}+A_{[6]}$$

$$\%6 := -A_{[4]}-A_{[5]}+A_{[6]}$$

```
> kirlv();
```

$$
\begin{bmatrix}
0 & 0 & 0 & -x_{[2]} & x_{[3]} & 0 \\
0 & 0 & 0 & x_{[1]} & 0 & -x_{[3]} \\
0 & 0 & 0 & 0 & -x_{[1]} & x_{[2]} \\
x_{[2]} & -x_{[1]} & 0 & 0 & -x_{[6]} & x_{[5]} \\
-x_{[3]} & 0 & x_{[1]} & x_{[6]} & 0 & -x_{[4]} \\
0 & x_{[3]} & -x_{[2]} & -x_{[5]} & x_{[4]} & 0
\end{bmatrix}
$$

```
> duals();
```

$$
\textit{xi-dagger}_{[1]} = \delta_{[1]}
$$

$$
\textit{xi-dagger}_{[2]} = \delta_{[2]}
$$

$$
\textit{xi-dagger}_{[3]} = \delta_{[3]}
$$

$$
\textit{xi-dagger}_{[4]} = -A_{[2]}\,\delta_{[1]} + A_{[1]}\,\delta_{[2]} + \delta_{[4]}
$$

$$
\textit{xi-dagger}_{[5]} = -\Big(-A_{[3]}\,\delta_{[1]}\cos\!\big(A_{[5]}\big) + A_{[1]}\,\delta_{[3]}\cos\!\big(A_{[5]}\big)
$$
$$
- \sin\!\big(A_{[4]}\big)\sin\!\big(A_{[5]}\big)\delta_{[4]} - \cos\!\big(A_{[4]}\big)\delta_{[5]}\cos\!\big(A_{[5]}\big) - \sin\!\big(A_{[4]}\big)\delta_{[6]}\Big)\Big/\Big(
$$
$$
\cos\!\big(A_{[5]}\big)\Big)
$$

$$
\textit{xi-dagger}_{[6]} = \Big(-A_{[3]}\,\delta_{[2]}\cos\!\big(A_{[5]}\big) + A_{[2]}\,\delta_{[3]}\cos\!\big(A_{[5]}\big)
$$
$$
+ \cos\!\big(A_{[4]}\big)\sin\!\big(A_{[5]}\big)\delta_{[4]} - \sin\!\big(A_{[4]}\big)\delta_{[5]}\cos\!\big(A_{[5]}\big) + \cos\!\big(A_{[4]}\big)\delta_{[6]}\Big)\Big/\Big(
$$
$$
\cos\!\big(A_{[5]}\big)\Big)
$$

$$
\textit{xi-star}_{[1]} = \cos\!\big(A_{[4]}\big)\cos\!\big(A_{[5]}\big)\delta_{[1]} + \sin\!\big(A_{[4]}\big)\cos\!\big(A_{[5]}\big)\delta_{[2]} - \sin\!\big(A_{[5]}\big)\delta_{[3]}
$$

$$
\textit{xi-star}_{[2]} = \delta_{[1]}\cos\!\big(A_{[4]}\big)\sin\!\big(A_{[5]}\big)\sin\!\big(A_{[6]}\big) - \delta_{[1]}\sin\!\big(A_{[4]}\big)\cos\!\big(A_{[6]}\big)
$$
$$
+ \delta_{[2]}\sin\!\big(A_{[4]}\big)\sin\!\big(A_{[5]}\big)\sin\!\big(A_{[6]}\big) + \delta_{[2]}\cos\!\big(A_{[4]}\big)\cos\!\big(A_{[6]}\big)
$$

$$+ \cos\left(A_{[5]}\right) \sin\left(A_{[6]}\right) \delta_{[3]}$$

$$\textit{xi-star}_{[3]} = \delta_{[1]} \cos\left(A_{[4]}\right) \sin\left(A_{[5]}\right) \cos\left(A_{[6]}\right) + \delta_{[1]} \sin\left(A_{[4]}\right) \sin\left(A_{[6]}\right)$$

$$+ \delta_{[2]} \sin\left(A_{[4]}\right) \sin\left(A_{[5]}\right) \cos\left(A_{[6]}\right) - \delta_{[2]} \cos\left(A_{[4]}\right) \sin\left(A_{[6]}\right)$$

$$+ \cos\left(A_{[5]}\right) \cos\left(A_{[6]}\right) \delta_{[3]}$$

$$\textit{xi-star}_{[4]} = \frac{\cos\left(A_{[6]}\right) \delta_{[4]} - \sin\left(A_{[6]}\right) \delta_{[5]} \cos\left(A_{[5]}\right) + \sin\left(A_{[5]}\right) \cos\left(A_{[6]}\right) \delta_{[6]}}{\cos\left(A_{[5]}\right)}$$

$$\textit{xi-star}_{[5]} = \frac{\sin\left(A_{[6]}\right) \delta_{[4]} + \cos\left(A_{[6]}\right) \delta_{[5]} \cos\left(A_{[5]}\right) + \sin\left(A_{[5]}\right) \sin\left(A_{[6]}\right) \delta_{[6]}}{\cos\left(A_{[5]}\right)}$$

$$\textit{xi-star}_{[6]} = \delta_{[6]}$$

$$\textit{xi-hat}_{[1]} = R_{[1]}$$

$$\textit{xi-hat}_{[2]} = R_{[2]}$$

$$\textit{xi-hat}_{[3]} = R_{[3]}$$

$$\textit{xi-hat}_{[4]} = -V_{[2]} R_{[1]} + V_{[1]} R_{[2]} + R_{[4]}$$

$$\textit{xi-hat}_{[5]} = -\left(-V_{[3]} R_{[1]} \cos\left(V_{[5]}\right) + V_{[1]} R_{[3]} \cos\left(V_{[5]}\right) \right.$$
$$\left. - \sin\left(V_{[4]}\right) \sin\left(V_{[5]}\right) R_{[4]} - \cos\left(V_{[4]}\right) R_{[5]} \cos\left(V_{[5]}\right) - \sin\left(V_{[4]}\right) R_{[6]} \right) \Big/ \Big($$
$$\cos\left(V_{[5]}\right) \Big)$$

$$\textit{xi-hat}_{[6]} = \left(-V_{[3]} R_{[2]} \cos\left(V_{[5]}\right) + V_{[2]} R_{[3]} \cos\left(V_{[5]}\right) \right.$$
$$\left. + \cos\left(V_{[4]}\right) \sin\left(V_{[5]}\right) R_{[4]} - \sin\left(V_{[4]}\right) R_{[5]} \cos\left(V_{[5]}\right) + \cos\left(V_{[4]}\right) R_{[6]} \right) \Big/ \Big($$
$$\cos\left(V_{[5]}\right) \Big)$$

This is su(2,2), Hermitian symmetric space of type I. The embedding is
into sl(4).

```
> X:=matrix(4,4,[[2*a[5]+a[6],a[8],a[1],a[2]],[a[9],2*a[5]-a[6],a[3],a[4]],[-a[15],-a[13],
> -2*a[5]+a[7],a[10]],[-a[14],-a[12],a[11],-2*a[5]-a[7]]]);
```

$$
X := \begin{bmatrix}
2\,a_{[5]} + a_{[6]} & a_{[8]} & a_{[1]} & a_{[2]} \\
a_{[9]} & 2\,a_{[5]} - a_{[6]} & a_{[3]} & a_{[4]} \\
-a_{[15]} & -a_{[13]} & -2\,a_{[5]} + a_{[7]} & a_{[10]} \\
-a_{[14]} & -a_{[12]} & a_{[11]} & -2\,a_{[5]} - a_{[7]}
\end{bmatrix}
$$

```
> d:=15:n:=4:lie(X,d,n,0);
```

left-dual

[continued]

> **indrep(4);**

$$inducedrep_{[1]} = R_{[1]}$$

$$inducedrep_{[2]} = R_{[2]}$$

$$inducedrep_{[3]} = R_{[3]}$$

$$inducedrep_{[4]} = R_{[4]}$$

$$inducedrep_{[5]} = 4\,V_{[1]}R_{[1]} + 4\,V_{[2]}R_{[2]} + 4\,V_{[3]}R_{[3]} + 4\,V_{[4]}R_{[4]} + t$$

$$inducedrep_{[6]} = V_{[1]}R_{[1]} + V_{[2]}R_{[2]} - V_{[3]}R_{[3]} - V_{[4]}R_{[4]}$$

$$inducedrep_{[7]} = -V_{[1]}R_{[1]} + V_{[2]}R_{[2]} - V_{[3]}R_{[3]} + V_{[4]}R_{[4]}$$

$$inducedrep_{[8]} = V_{[3]}R_{[1]} + V_{[4]}R_{[2]}$$

$$inducedrep_{[9]} = V_{[1]}R_{[3]} + V_{[2]}R_{[4]}$$

$$inducedrep_{[10]} = -V_{[1]}R_{[2]} - V_{[3]}R_{[4]}$$

$$inducedrep_{[11]} = -V_{[2]}R_{[1]} - V_{[4]}R_{[3]}$$

$$inducedrep_{[12]} =$$

$$V_{[2]}V_{[3]}R_{[1]} + V_{[2]}V_{[4]}R_{[2]} + V_{[3]}V_{[4]}R_{[3]} + V_{[4]}^{2}R_{[4]} + \frac{1}{4}V_{[4]}t$$

$$inducedrep_{[13]} =$$

$$V_{[1]}V_{[3]}R_{[1]} + V_{[1]}V_{[4]}R_{[2]} + V_{[3]}^{2}R_{[3]} + V_{[3]}V_{[4]}R_{[4]} + \frac{1}{4}V_{[3]}t$$

$$inducedrep_{[14]} =$$

$$V_{[1]}V_{[2]}R_{[1]} + V_{[2]}^{2}R_{[2]} + V_{[1]}V_{[4]}R_{[3]} + V_{[2]}V_{[4]}R_{[4]} + \frac{1}{4}V_{[2]}t$$

$$inducedrep_{[15]} =$$

$$V_{[1]}^{2}R_{[1]} + V_{[1]}V_{[2]}R_{[2]} + V_{[1]}V_{[3]}R_{[3]} + V_{[2]}V_{[3]}R_{[4]} + \frac{1}{4}V_{[1]}t$$

This is so(6) with the space of 3-by-3 skew-symmetric matrices embedded in the upper-right corner.

```
> X:=matrix(6,6,[a[4]+2*a[6],a[7],a[8],0,a[3],-a[2],a[10],a[4]+a[5]-a[6],a[9],-a[3],0,a[1],
> a[11],a[12],a[4]-a[5]-a[6],a[2],-a[1],0,0,a[13],-a[14],-a[4]-2*a[6],-a[10],-a[11],-a[13],0
> ,a[15],-a[7],-a[4]-a[5]+a[6],-a[12],a[14],-a[15],0,-a[8],-a[9],-a[4]+a[5]+a[6]]);
```

$$X := \begin{bmatrix} a_{[4]}+2\,a_{[6]}, & a_{[7]}, & a_{[8]}, & 0, & a_{[3]}, & -a_{[2]} \\ a_{[10]}, & a_{[4]}+a_{[5]}-a_{[6]}, & a_{[9]}, & -a_{[3]}, & 0, & a_{[1]} \\ a_{[11]}, & a_{[12]}, & a_{[4]}-a_{[5]}-a_{[6]}, & a_{[2]}, & -a_{[1]}, & 0 \\ 0, & a_{[13]}, & -a_{[14]}, & -a_{[4]}-2\,a_{[6]}, & -a_{[10]}, & -a_{[11]} \\ -a_{[13]}, & 0, & a_{[15]}, & -a_{[7]}, & -a_{[4]}-a_{[5]}+a_{[6]}, & -a_{[12]} \\ a_{[14]}, & -a_{[15]}, & 0, & -a_{[8]}, & -a_{[9]}, & -a_{[4]}+a_{[5]}+a_{[6]} \end{bmatrix}$$

```
> d:=15:n:=6:lie(X,d,n,0);
```

left-dual

[continued]

> indrep(3);

$$inducedrep_{[1]} = R_{[1]}$$

$$inducedrep_{[2]} = R_{[2]}$$

$$inducedrep_{[3]} = R_{[3]}$$

$$inducedrep_{[4]} = 2\,V_{[1]}\,R_{[1]} + 2\,V_{[2]}\,R_{[2]} + 2\,V_{[3]}\,R_{[3]} + t$$

$$inducedrep_{[5]} = -V_{[2]}\,R_{[2]} + V_{[3]}\,R_{[3]}$$

$$inducedrep_{[6]} = -2\,V_{[1]}\,R_{[1]} + V_{[2]}\,R_{[2]} + V_{[3]}\,R_{[3]}$$

$$inducedrep_{[7]} = -V_{[1]}\,R_{[2]}$$

$$inducedrep_{[8]} = -V_{[1]}\,R_{[3]}$$

$$inducedrep_{[9]} = -V_{[2]}\,R_{[3]}$$

$$inducedrep_{[10]} = -V_{[2]}\,R_{[1]}$$

$$inducedrep_{[11]} = -V_{[3]}\,R_{[1]}$$

$$inducedrep_{[12]} = -V_{[3]}\,R_{[2]}$$

$$inducedrep_{[13]} = V_{[2]}\,V_{[3]}\,R_{[2]} + V_{[1]}\,V_{[3]}\,R_{[1]} + \frac{2}{3}\,V_{[3]}\,t + V_{[3]}^{\,2}\,R_{[3]}$$

$$inducedrep_{[14]} = V_{[1]}\,V_{[2]}\,R_{[1]} + V_{[2]}^{\,2}\,R_{[2]} + V_{[2]}\,V_{[3]}\,R_{[3]} + \frac{2}{3}\,V_{[2]}\,t$$

$$inducedrep_{[15]} = V_{[1]}^{\,2}\,R_{[1]} + V_{[1]}\,V_{[2]}\,R_{[2]} + V_{[1]}\,V_{[3]}\,R_{[3]} + \frac{2}{3}\,V_{[1]}\,t$$

This is sp(4). The corresponding homogeneous space of symmetric 2x2 matrices is embedded in the upper right corner.

```
> X:=matrix(4,4,[[a[4]+a[5],a[6],a[1],a[2]],[a[7],a[4]-a[5],a[2],a[3]],[-a[8],-a[9],
> -a[4]-a[5],-a[7] ],[-a[9],-a[10],-a[6], -a[4]+a[5]]]);
```

$$X := \begin{bmatrix} a_{[4]} + a_{[5]} & a_{[6]} & a_{[1]} & a_{[2]} \\ a_{[7]} & a_{[4]} - a_{[5]} & a_{[2]} & a_{[3]} \\ -a_{[8]} & -a_{[9]} & -a_{[4]} - a_{[5]} & -a_{[7]} \\ -a_{[9]} & -a_{[10]} & -a_{[6]} & -a_{[4]} + a_{[5]} \end{bmatrix}$$

```
> d:=10:n:=4:lie(X,d,n,0);
```

left-dual

[continued]

```
> kirlv();
```

$$
\begin{bmatrix}
0,0,0,-2x_{[1]},-2x_{[1]},0,-x_{[2]},-\frac{1}{2}x_{[4]}-\frac{1}{2}x_{[5]},-x_{[6]},0 \\[2mm]
0,0,0,-2x_{[2]},0,-2x_{[1]},-2x_{[3]},-x_{[7]},-x_{[4]},-x_{[6]} \\[2mm]
0,0,0,-2x_{[3]},2x_{[3]},-x_{[2]},0,0,-x_{[7]},-\frac{1}{2}x_{[4]}+\frac{1}{2}x_{[5]} \\[2mm]
2x_{[1]},2x_{[2]},2x_{[3]},0,0,0,0,-2x_{[8]},-2x_{[9]},-2x_{[10]} \\[2mm]
2x_{[1]},0,-2x_{[3]},0,0,2x_{[6]},-2x_{[7]},-2x_{[8]},0,2x_{[10]} \\[2mm]
0,2x_{[1]},x_{[2]},0,-2x_{[6]},0,x_{[5]},-x_{[9]},-2x_{[10]},0 \\[2mm]
x_{[2]},2x_{[3]},0,0,2x_{[7]},-x_{[5]},0,0,-2x_{[8]},-x_{[9]} \\[2mm]
\frac{1}{2}x_{[4]}+\frac{1}{2}x_{[5]},x_{[7]},0,2x_{[8]},2x_{[8]},x_{[9]},0,0,0,0 \\[2mm]
x_{[6]},x_{[4]},x_{[7]},2x_{[9]},0,2x_{[10]},2x_{[8]},0,0,0 \\[2mm]
0,x_{[6]},\frac{1}{2}x_{[4]}-\frac{1}{2}x_{[5]},2x_{[10]},-2x_{[10]},0,x_{[9]},0,0,0
\end{bmatrix}
$$

[continued]

> indrep(3);

$$inducedrep_{[1]} = R_{[1]}$$

$$inducedrep_{[2]} = R_{[2]}$$

$$inducedrep_{[3]} = R_{[3]}$$

$$inducedrep_{[4]} = 2\,V_{[1]}\,R_{[1]} + 2\,V_{[2]}\,R_{[2]} + 2\,V_{[3]}\,R_{[3]} + t$$

$$inducedrep_{[5]} = 2\,V_{[1]}\,R_{[1]} - 2\,V_{[3]}\,R_{[3]}$$

$$inducedrep_{[6]} = 2\,V_{[2]}\,R_{[1]} + V_{[3]}\,R_{[2]}$$

$$inducedrep_{[7]} = V_{[1]}\,R_{[2]} + 2\,V_{[2]}\,R_{[3]}$$

$$inducedrep_{[8]} = V_{[1]}^{2}\,R_{[1]} + V_{[2]}\,V_{[1]}\,R_{[2]} + V_{[2]}^{2}\,R_{[3]} + \frac{1}{2}\,V_{[1]}\,t$$

$$inducedrep_{[9]} =$$

$$2\,V_{[2]}\,V_{[1]}\,R_{[1]} + R_{[2]}\,V_{[2]}^{2} + R_{[2]}\,V_{[3]}\,V_{[1]} + 2\,V_{[3]}\,V_{[2]}\,R_{[3]} + V_{[2]}\,t$$

$$inducedrep_{[10]} = V_{[2]}^{2}\,R_{[1]} + V_{[3]}\,V_{[2]}\,R_{[2]} + V_{[3]}^{2}\,R_{[3]} + \frac{1}{2}\,V_{[3]}\,t$$

LOMMEL DISTRIBUTION FOR t=70

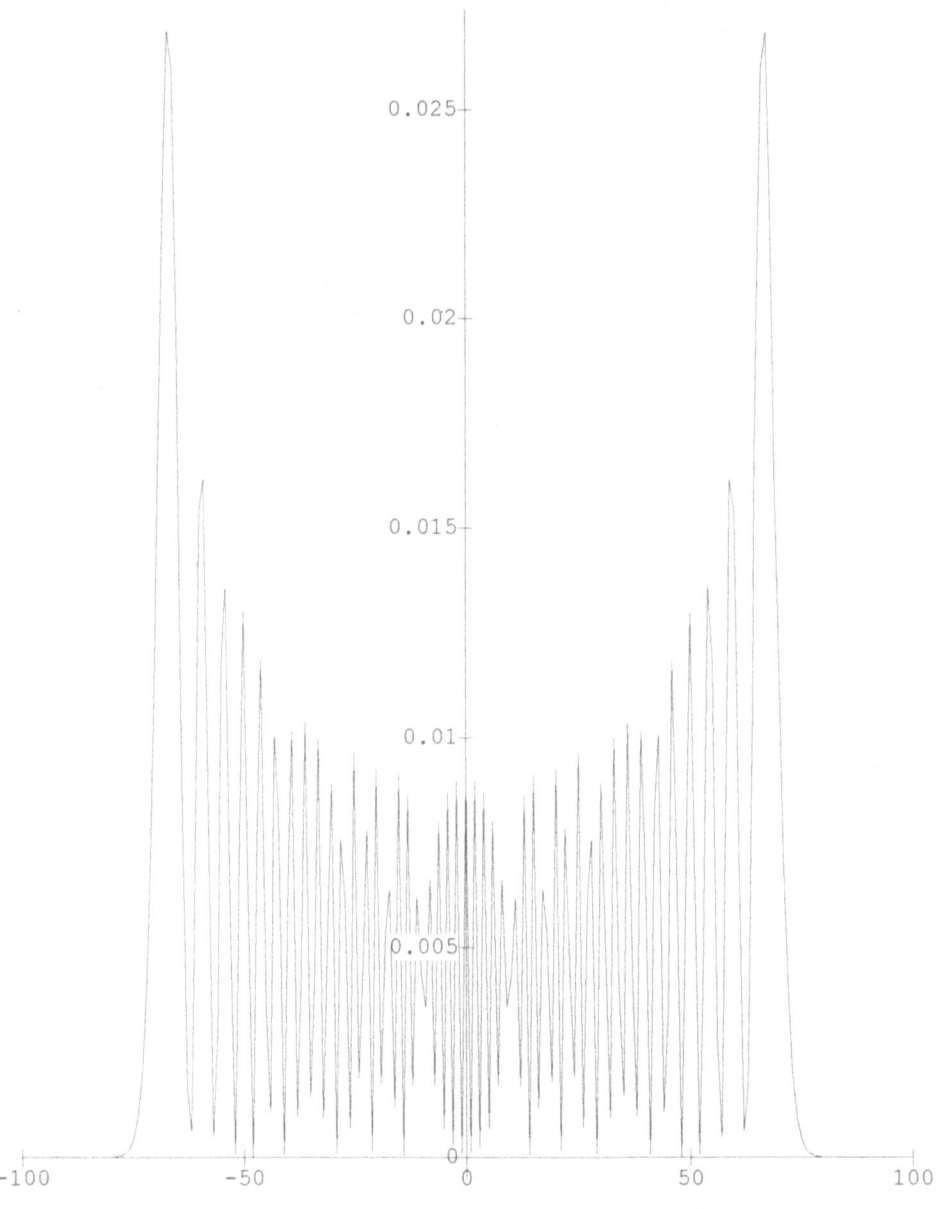

NOTE: *A comment followed by *L means that the procedure Lie should be run first.*
Otherwise, the procedure can be used anywhere.
with(linalg):

#lie1: *This is the core procedure*

```
lie1 := proc (XY:matrix, dd:integer, n:integer,tt:integer)
local i,k,kk;
m := n^2;
Adot := 'Adot';
for i to dd do xi[i] := map(diff,XY,a[i]) od;
gg := multiply(seq( exponential(scalarmul(xi[kk],A[kk])),kk
= 1 .. dd));
gg:=map(simplify,gg);
#TO SHORTEN CALCULATIONS AND DISPLAYS, # OUT THE NEXT LINE
hh:=gg;
gg:=map(convert,gg,exp);gg:=map(expand,gg);
#print(`the group element is computed`);
Xg := multiply(XY,gg) ;
for i to dd do ggdot[i] := Adot[i]*map(diff,gg,A[i]) od;
gdot := evalm(sum(ggdot[k],k = 1 .. dd));
Xgv := convert(Xg,vector);
gdotv := convert(gdot,vector);
gdotv := map(convert,gdotv,exp);
Xgv := map(convert,Xgv,exp);
#print(`solving equations for pi-dagger`);
sys:=solve({seq(gdotv[i]-Xgv[i],i = 1 .. m)},{seq(Adot[i],i
= 1 .. dd)});
assign(sys); vv := convert([seq(Adot[i],i = 1 ..
dd)],vector);
for i to dd do pdgr[i] := map(diff,vv,a[i]) od;
pi[1] := concat(seq(pdgr[k],k = 1 .. dd));
Adot:='Adot';
gX := multiply(gg,XY); gXv := convert(gX,vector);
gXv := map(convert,gXv,exp);
#print(`solving equations for pi-star`);
sys := solve({seq(gdotv[i]-gXv[i],i = 1 ..
m)},{seq(Adot[i],i = 1 .. dd)});
assign(sys);
vv := convert([seq(Adot[i],i = 1 .. dd)],vector);
for i to dd do pst[i] := map(diff,vv,a[i]) od;
pi[2] := concat(seq(pst[k],k = 1 .. dd));
pi[1]:=map(simplify,pi[1]);pi[2]:=map(simplify,pi[2]);
if tt=1 then trg() fi;
#print(`computing piadj`);
pidgr := transpose(pi[1]); pistar := transpose(pi[2]);
#TO SHORTEN CALCULATIONS AND DISPLAYS, # OUT THE NEXT LINE
piadj :=map(simplify,evalm(`&*`(1/pi[1],pi[2])));
hh,piadj, pidgr, pistar
 end:
```

#see: *Prints results of Lie1 (part of Lie)*
```
see := proc () local i, prt;
prt := [`group element`,`exp-adjoint`, `left-dual`, `right-
dual`];
for i to 4 do print(prt[i]); print(); print(pimat[i]);
print() od
end:
```

#trg: *Converts output to trigonometric form if desired (part of Lie)*
```
trg:=proc() local ix;
for ix to 2
do:pi[ix]:=map(simplify,map(convert,map(expand,pi[ix]),trig)
);od;
end:
```

#lie: *Calls the core procedures*
```
lie := proc (Y:matrix,ddd:integer,nn:integer,tt:integer)
        local i;
        pimat := lie1(Y,ddd,nn,tt);
        see()
        end:
lie;
```

#cmm: *Calculates commutator of two matrices*
```
cmm:=proc(XX,YY)
evalm(XX&*YY-YY&*XX)
end:
cmm;
```

#kll: *Calculates the trace of the product of two matrices*
```
kll:=proc(XX,YY)
local hhh,ZZ1,ZZ2;
    ZZ1 := convert(XX,vector);
    ZZ2 := convert(YY,vector);
    hhh := dotprod(ZZ1,ZZ2);
    op(hhh)
end:
kll;
```

#cffs: *Calculates the coefficients of the expansion of a matrix in an orthogonal basis*

```
cffs:=proc(QX,y,dd)
local hhh,yy,YY;
    YY := convert(QX,vector);
    for i to dd do  yy[i] := convert(y[i],vector) od;
    for i to dd do
        hhh[i] := dotprod(yy[i],YY)/dotprod(yy[i],yy[i])
    od;
    op(hhh)
end:
cffs;
```

#liealg: *Given the general element X of the Lie algebra, produces the basis*

```
liealg:=proc(XY)
for i to d do xi[i] := map(diff,XY,a[i]) od;
op(xi)
end:
liealg;
```

#kirlv: *Calculates the Kirillov form (*L)*

```
kirlv := proc()
local i, ii, ix;
xvec := matrix(1,d,proc (i, j) options operator, arrow; x[j]
end);
picheck:=transpose(pistar-pidgr);
for i to d do:adjt[i]:=map(eval,subs({ seq(A[ix] = 0,ix = 1
.. d)},
map(diff,evalm(picheck),A[i])))):od;
for i to d do:
krow[i] := multiply(xvec,adjt[i]): od;
KRLVA := stack(seq(krow[ii],ii = 1 .. d));
KRLV := map(eval,KRLVA);
print(KRLV)
end:
kirlv;
```

#matrec: *Calculates the recurrence for the matrix elements (*L)*

```
matrec:=proc()
local i;
pdv:=subs(A=V,evalm(pidgr));
pdrr:=subs(A=RR,evalm(pidgr));
VV0:=convert(vector([seq(VV[i],i=1..d)]),matrix);
Ronn:=map(simplify,multiply(pdv^(-1),
transpose(piadj),pdrr,VV0));
map(expand,Ronn);
end:
matrec;
```

#grp: *Calculates the coefficients of the product of group elements
with respect to the basis xi (*L)*

```
grp := proc () local h, i, yt;
gg := multiply(seq( exponential(scalarmul(xi[kk],A[kk])),kk
= 1 .. d));
gg:=map(simplify,gg);
g1g2 := evalm(`&*`(gg,subs(A = B,evalm( gg))));
hv := convert(g1g2,vector);
for i to d do yt := convert(xi[i], vector);
AXB[i] := dotprod(hv,yt)/dotprod(yt,yt) od;
print(group-entries),
print();
eval(AXB) end:
grp;
```

#grpels: *After running* **grp**, *solves for the group law (*L)*

```
grpels := proc ()
Y:='Y';
grx := subs(A = Y,evalm(hh));
grxv := convert(cffs(grx,xi,d),array);
AXBv := convert(AXB,array);
grpsys := solve({seq(grxv[i] = AXB[i],i = 1 ..
d)},{seq(Y[i],i = 1 .. d)});
assign(grpsys);
grpel := map(simplify,eval(Y));
for i to d do print(`groupel`[i]=grpel[i]) od
end:
grpels;
```

#duals: *Calculates the dual representations, xi-dagger, xi-star, and*
        xi-hat (*L)

```
duals:=proc()
delta:=vector(d);
xidagger:=multiply(pidgr,delta);
xistar:=multiply(pistar,delta);
for i to d do:print(`xi-dagger`[i]=xidagger[i]);od;
print(` `);
print(`----------------------------------------------------
`);
print(` `);
for i to d do:print(`xi-star`[i]=xistar[i]);od;
print(` `);
print(`----------------------------------------------------
`);
print(` `);
dbldual:=subs({delta=R,A=V},evalm(xidagger));
for i to d do:print(`xi-hat`[i]=dbldual[i]);od;
end:
duals;
```


#indrep: *Calculates the induced representation, mapping the next*
        basis element after the R's to the scalar t, and setting the
        remaining R's to zero (*L)

```
indrep:=proc(N:integer)
induced:=subs({seq(R[j]=0,j-N+2..d),R[N+1]=t},evalm(dbldual)
);
for i to d do:print(`inducedrep`[i]=induced[i]);od;
end:
indrep;
```


#adjrep: *After running* **kirlv,** *call this to print out the matrices of the*
        adjoint representation (*L)

```
adjrep:=proc()
for i to d do:print(`adjrep`[i]=evalm(adjt[i])):od
end:
adjrep;
```

References

1. M. Abramowitz and I. Stegun, *Handbook of mathematical functions*, US Govt. Printing Office, 1972.

2. F. Berezin, *Quantization in complex symmetric spaces*, Math. USSR Izvestia, **9** (1975), 341–379.

3. L. Biedenharn and L. Louck, *Angular momentum in quantum theory*, Addison-Wesley, 1981.

4. P. Bougerol, *Comportement asymptotique des puissances de convolution d'une probabilité sur un espace symétrique*, Astérisque, **74**, 1980, 29–46.

5. P. Bougerol, *Théorème central limite sur certains groupes de Lie*, Annales Sci. de l'ENS, **14**, 1980, 403-432.

6. N. Bourbaki, *Algèbres et groupes de Lie*, Chaps. 1–3, CCLS, 1971-72.

7. D. Cheng, W.P. Dayawansa and C.F. Martin, *Observability of systems on Lie groups and coset spaces*, SIAM J. Control and Optimisation, **28**, 3, 1990, 570–581.

8. C. Chevalley, *Theory of Lie groups*, Princeton University Press, 1946.

9. P.E. Crouch and M. Irving, *On finite Volterra series which admit hamiltonian realizations*, Math. Systems Theory **17**, 1984, 293–318.

10. J. Dixmier, *Enveloping algebras*, [translated by Minerva Translations, ltd., London]. Amsterdam: North-Holland Pub. Co., 1977.

11. G. Duchamp and D. Krob, *The partially commutative free Lie algebra: bases and ranks*, Adv. in Math., **95**, 1992, 92–126.

12. G. Duchamp and D. Krob, *Computing with P.B.W. in enveloping algebras*, report LITP 91.11, 1991.

13. P. Feinsilver, U. Franz, and R. Schott, *On the computation of polynomial representations of nilpotent Lie groups: a symbolic mathematical approach*, Institut E. Cartan, Rapport de Recherche **27**, 1994.

14. P. Feinsilver, U. Franz, and R. Schott, *On duality and stochastic processes on quantum groups*, Conference Proceedings, Intl. Symp. on Nonlinear, Dissipative, Irreversible Quantum Systems, Aug., 1994, 29-36, World Scientific Press.

15. P. Feinsilver, J. Kocik, and R. Schott, *Representations and stochastic processes on groups of type-H*, J. Funct. Analysis., **115**, 1, 1993, 146–165.

16. P. Feinsilver and R. Schott, *Computing representations of a Lie group via the universal enveloping algebra*, Institut E. Cartan, Rapport de Recherche **7**, 1995.

17. P. Feinsilver and R. Schott, *Differential relations and recurrence formulas for representations of Lie groups*, Studies in Applied Mathematics (in press).

18. P. Feinsilver and R. Schott, *Algebraic structures and operator calculus, vol. 2: Special functions and computer science*, Kluwer Academic Publishers, 1994.

19. P. Feinsilver and R. Schott, *Algebraic structures and operator calculus, vol. 1: Representations and probability theory*, Kluwer Academic Publishers, 1993.

20. P. Feinsilver and R. Schott, *Operator calculus approach to orthogonal polynomial expansions*, Rapport INRIA **1745**, 1992 and Journal of Computational and Applied Mathematics (in press).

21. P. Feinsilver and R. Schott, *On Bessel functions and rate of convergence of zeros of Lommel polynomials*, Math. Comp., **59**, 199, 1992, 153–156.

22. P. Feinsilver and R. Schott, *Appell systems on Lie groups*, J. Theo. Prob., **5** (1992), 251–281.

23. P. Feinsilver and R. Schott, *Krawtchouk polynomials and finite probability theory*, Conference Proceedings: Probability Measures on Groups X, 129–136, Plenum Press, 1991.

24. P. Feinsilver and R. Schott, *Special functions and infinite-dimensional representations of Lie groups*, Math. Zeit., **203**, 1990, 173–191.

25. P. Feinsilver and R. Schott, *An operator approach to processes on Lie groups*, Lecture Notes in Mathematics, Vol. **1391**, 59-65, Conference Proceedings, Probability Theory on Vector Spaces, Lancut 1987, Springer-Verlag, 1989.

26. P. Feinsilver and R. Schott, *Operators, stochastic processes, and Lie groups*, Lecture Notes in Mathematics, Vol. **1379**, 75-85, Oberwolfach Conference, Probability Measures on Groups, Springer-Verlag, 1989.

27. M. Fliess, *Fonctionnelles causales non linéaires et indéterminées non commutatives*, Bull. Soc. Math. France, **109**, 1981, 3–40.

28. M. Fliess and D. Normand-Cyrot, *Algèbres de Lie nilpotentes, formule de Baker-Campbell-Hausdorff et intègrales itérées de K.T. Chen*, Séminaire de Probabilités, XVI, Lect. Notes in Math., Springer-Verlag, 1982.

29. R.D. Gill, *Lectures on survival analysis*, St. Flour Lectures on Probability Theory, Lect. Notes in Math., **1581**, 1994, 115–236.

30. R.D. Gill and S. Johansen, *A survey of product integration with a view toward applications in survival analysis*, Ann. Stat., **18**, 4, 1990, 1501–1555.

31. B. Gruber and A.U. Klimyk, *Indecomposable representations of su(2)*, J. Math. Phys., **15**, 1986, 201–236.

32. K.T. Hecht, *The vector coherent state method and its application to problems of higher symmetries*, Springer Lecture Notes in Physics, **290**, 1987.

33. H. Hermes, *Nilpotent and high-order approximations of vector field systems*, SIAM REVIEW, **33**, 2, 1991, 238–264.

34. H. Heyer, *Probabiliy measures on locally compact groups*, Springer Verlag, 1979.

35. L.K. Hua, *Harmonic analysis of functions of several complex variables in the classical domains*, Transl. Math. Mono., **6**, Amer. Math. Soc., Providence, 1963.

36. N. Hurt, *Geometric quantization in action: applications of harmonic analysis in quantum statistical mechanics and quantum field theory*, Kluwer, 1983.

37. H.P. Jakobsen, *Hermitian symmetric spaces and their unitary highest weight modules*, J. Funct. Analysis, **52**, 1983, 385–412.

38. B. Jakubczyk and E.D. Sontag, *Controllability of nonlinear discrete-time systems: a Lie-algebraic approach*, SIAM J. Control and Optim., **28**, 1, 1990, 1–33.

39. J.R. Klauder and B.S. Skagerstam, *Coherent states*, applications in physics and mathematical physics, World Scientific Publishing, 1985.

40. A.U. Klimyk & N. Ya. Vilenkin, *Representation of Lie groups and special functions*, (4 vols), Kluwer Academic Publishers, 1991-1994.

41. R. Koekoek and R.F. Swarttouw, *The Askey-scheme of hypergeometric orthogonal polynomials and its q-analogue*, Report **94-05**, Tech. Univ. Delft, 1994.

42. T.H. Koornwinder, *Handling hypergeometric series in MAPLE*, in Orthogonal polynomials and their applications, IMACS Annals on Comp. and Appl. Math. **9**, Baltzer, 1991, 73–80.

43. T.H. Koornwinder & I. Sprinkhuizen-Kuyper, *Hypergeometric functions of 2×2 matrix argument are expressible in terms of Appell's function F_4*, Proc. AMS., **70**, 1978, 39–42.

44. M. A. A. van Leeuwen, A. M. Cohen and B. Lisser, *LIE*, A Package for Lie Group Computations, Published by CAN, 1992.

45. M. Lorente, *Unitarity of representations on Verma modules*, Symmetries in Science V, Plenum Press, 1990.

46. R.M. Murray, Z. Li, S.S. Sastry, *A mathematical introduction to robotic manipulation*, CRC Press, 1994.

47. A. Perelomov, *Generalized coherent states and applications*, Springer-Verlag, 1986.

48. I. Satake, *Algebraic structures of symmetric domains*, Iwanami Shoten and Princeton University Press, 1980.

49. H.J. Sussmann, *A general theorem on local controllability*, SIAM J. Control Optim., **25**, 1987, 158–194.

50. M.V. Tratnik, *Multivariable continuous Hahn polynomials*, J. Math. Phys., **29**, 1988, 1529ff.

51. X.G. Viennot, *Algèbres de Lie libres et monoïdes libres*, Lect. Notes in Math., **691**, Springer-Verlag, 1978.

52. G.N. Watson, *Theory of Bessel functions*, Cambridge U. Press, 1980.

53. J. Wei and E. Norman, *On global representation of the solutions of linear differential equations as a product of exponentials*, Proc. A.M.S., **15**, 1964, 327–334.

54. J. Wei and E. Norman, *Lie algebraic solution of linear differential equations*, J. Math. Phys., **4**, 1963, 575–581.

55. J.A. Wolf, *Fine structure of Hermitian symmetric spaces*, in *Symmetric Spaces*, W.M. Boothby & G.L. Weiss (eds.), Marcel Dekker, New York, 1972, 271–357.

56. N. Woodhouse, *Geometric quantization*, Oxford University Press, 2nd edition, 1992.

57. W.-M. Zhang, D.H. Feng, and R. Gilmore, *Coherent states: Theory and some applications*, Rev. Mod. Phys, **62**, 4, 1990, 867–928.

INDEX

θ-polynomials, 22

ABCD-type algebras, 126

abelian, 9

addition formula, 29

adjoint action, 54

adjoint group, 15, 36

adjoint group and coadjoint orbits, 14

adjoint orbits, 15

adjoint representation, 9

Ado's Theorem, 1

affine algebra, 127

affine type, 10

affine type algebras, 31

anti-commutation, 139

Appell duality, 20

Appell systems, 6

associativity construction, 81

basic hypergeometric function, 92

Bessel functions, 109

Lommel polynomials, 108

Berezin symbol, 62

Bernoulli system, 79

Bessel functions, 106

Bessel polynomials, 22

binomial coefficient matrix and transposition symmetry, 96

boson calculus, 3, 6

boson operators, 17

Brownian motion, 49

canonical Appell systems, 66, 146

canonical basis polynomials, 20

canonical coordinates, 19

canonical dual, 19

canonical systems, 26

Cartan decomposition, 137, 143

Cartan's criteria, 10

Casimir operator, 102

center, 9

central series, 10

coadjoint orbits, 16, 37

cocycle property, 63

coherent state representation, 62

coherent states, 134, 141, 145

complete positivity, 63

construction of the algebra from coherent states, 65

convolutions and orthogonal functions, 78

cyclic vector, 2

definitions, 118

degree operator, 17

derived algebra, 10

derived series, 10

differential recurrences, 39

Other *Mathematics and Its Applications* titles of interest:

M.A. Frumkin: *Systolic Computations.* 1992, 320 pp.   ISBN 0-7923-1708-4

J. Alajbegovic and J. Mockor: *Approximation Theorems in Commutative Algebra.* 1992, 330 pp.   ISBN 0-7923-1948-6

I.A. Faradzev, A.A. Ivanov, M.M. Klin and A.J. Woldar: *Investigations in Algebraic Theory of Combinatorial Objects.* 1993, 516 pp.   ISBN 0-7923-1927-3

I.E. Shparlinski: *Computational and Algorithmic Problems in Finite Fields.* 1992, 266 pp.   ISBN 0-7923-2057-3

P. Feinsilver and R. Schott: *Algebraic Structures and Operator Calculus.* Vol. I. Representations and Probability Theory. 1993, 224 pp.   ISBN 0-7923-2116-2

A.G. Pinus: *Boolean Constructions in Universal Algebras.* 1993, 350 pp.
ISBN 0-7923-2117-0

V.V. Alexandrov and N.D. Gorsky: *Image Representation and Processing. A Recursive Approach.* 1993, 200 pp.   ISBN 0-7923-2136-7

L.A. Bokut' and G.P. Kukin: *Algorithmic and Combinatorial Algebra.* 1994, 384 pp.   ISBN 0-7923-2313-0

Y. Bahturin: *Basic Structures of Modern Algebra.* 1993, 419 pp.
ISBN 0-7923-2459-5

R. Krichevsky: *Universal Compression and Retrieval.* 1994, 219 pp.
ISBN 0-7923-2672-5

A. Elduque and H.C. Myung: *Mutations of Alternative Algebras.* 1994, 226 pp.
ISBN 0-7923-2735-7

E. Goles and S. Martínez (eds.): *Cellular Automata, Dynamical Systems and Neural Networks.* 1994, 189 pp.   ISBN 0-7923-2772-1

A.G. Kusraev and S.S. Kutateladze: *Nonstandard Methods of Analysis.* 1994, 444 pp.   ISBN 0-7923-2892-2

P. Feinsilver and R. Schott: *Algebraic Structures and Operator Calculus.* Vol. II. Special Functions and Computer Science. 1994, 148 pp.   ISBN 0-7923-2921-X

V.M. Kopytov and N. Ya. Medvedev: *The Theory of Lattice-Ordered Groups.* 1994, 400 pp.   ISBN 0-7923-3169-9

H. Inassaridze: *Algebraic K-Theory.* 1995, 438 pp.   ISBN 0-7923-3185-0

C. Mortensen: *Inconsistent Mathematics.* 1995, 155 pp.   ISBN 0-7923-3186-9

R. Abłamowicz and P. Lounesto (eds.): *Clifford Algebras and Spinor Structures.* A Special Volume Dedicated to the Memory of Albert Crumeyrolle (1919–1992). 1995, 421 pp.   ISBN 0-7923-3366-7

W. Bosma and A. van der Poorten (eds.), *Computational Algebra and Number Theory.* 1995, 336 pp.   ISBN 0-7923-3501-5

Other *Mathematics and Its Applications* titles of interest:

A.L. Rosenberg: *Noncommutative Algebraic Geometry and Representations of Quantized Algebras*. 1995, 316 pp. ISBN 0-7923-3575-9

L. Yanpei: *Embeddability in Graphs*. 1995, 400 pp. ISBN 0-7923-3648-8

B.S. Stechkin and V.I. Baranov: *Extremal Combinatorial Problems and Their Applications*. 1995, 205 pp. ISBN 0-7923-3631-3

Y. Fong, H.E. Bell, W.-F. Ke, G. Mason and G. Pilz (eds.): *Near-Rings and Near-Fields*. 1995, 278 pp. ISBN 0-7923-3635-6

A. Facchini and C. Menini (eds.): *Abelian Groups and Modules*. (Proceedings of the Padova Conference, Padova, Italy, June 23–July 1, 1994). 1995, 537 pp.
ISBN 0-7923-3756-5

D. Dikranjan and W. Tholen: *Categorical Structure of Closure Operators*. With Applications to Topology, Algebra and Discrete Mathematics. 1995, 376 pp.
ISBN 0-7923-3772-7

A.D. Korshunov (ed.): *Discrete Analysis and Operations Research*. 1996, 351 pp.
ISBN 0-7923-3866-9

P. Feinsilver and R. Schott: *Algebraic Structures and Operator Calculus*. Vol. III: Representations of Lie Groups. 1996, 238 pp. ISBN 0-7923-3834-0